Sports Analytics

A. Mansurali • P. Mary Jeyanthi
Dieu Hack-Polay • Ali B. Mahmoud
Editors

Sports Analytics

Data-Driven Sports and Decision Intelligence

Springer

Editors
A. Mansurali
Department of Management,
Central University of Tamil Nadu
Thiruvarur, Tamil Nadu, India

Dieu Hack-Polay
Crandall University
Moncton, NB, Canada

P. Mary Jeyanthi
Jaipuria Institute of Management
Jaipur, Rajasthan, India

Ali B. Mahmoud
St. John's University
New York, NY, USA

University of London
London, UK

ISBN 978-3-031-63572-4 ISBN 978-3-031-63573-1 (eBook)
https://doi.org/10.1007/978-3-031-63573-1

This Springer imprint is published by the registered company Springer Nature Switzerland AG
The registered company address is: Gewerbestrasse 11, 6330 Cham, Switzerland

Preface

The application of data analysis techniques to different aspects of sports, such as players' performance, recruitment, etc. has significantly increased in the last decade. The main aim is to aid strategic decision-making by stakeholders in the sports industry. This is what is increasingly referred to as sports analytics and termed on-field and off-field analytics. As a source of competitive advantage, sports analytics is slowly rising in prominence. In an era of intense competition, where every second on the field counts and decisions made can have substantial financial and career impacts, data is becoming the new currency that sports teams and players deploy. With the availability of sophisticated data collection techniques, more and more emphasis is placed on the maximization of data value by analyzing and predicting key performance indicators that will help attain maximum performance.

The use of machine learning techniques helps to augur on-field and off-field performance. This tangible effectiveness of machine learning has opened up new careers in sports analytics with top salaries. As a strategic tool, sports analytics can help create a sports ecosystem that would result in an end-to-end solution and improved performance.

The book has compiled interesting chapters covering various emerging research areas in sports analytics, e.g., envisaging in sports analytics, decision intelligence in sports marketing, sports talent identification and acquisition, IoT and blockchain in sports, sports revenue analytics, sports marketing analytics, forecasting in sport, classification and regression techniques for sports, and sports data collection and consumption methods. We found this coverage significant for many different reasons.

In fact, with the quantum of data going up in sports analytics, data visualization is gaining prominence. From basic sports dashboards to advanced infographics, visualization techniques can present data in an easily comprehendible and space-effective way. At the same time, sports marketing and sponsorship are becoming highly impactful areas for determining success as the market becomes more competitive. It is accepted that this allows marketers to capitalize on the popularity and attention of sports to promote a brand or product, thus increasing revenue. In the same vein, the search for talent is the subject of fierce competition. In this field, too, data analytics use is playing a vital role, helping teams source the most promising athletes. By analyzing data from a range of sources, including performance metrics, scouting reports, and social media activity, teams can identify patterns and trends that are indicative of potential success in a particular sport or position. This information can be utilized to build a more complete and accurate picture of a player's skills, abilities, and potential and to identify players who may have been bypassed by traditional scouting approaches. Data analytics can also be used to track the progress and development of players over time, identifying areas for improvement and tailoring training programs to meet individual needs. By leveraging the power of data analytics, sports teams can make more informed and strategic decisions about talent identification and acquisition, ultimately leading to a more successful and competitive team.

The use of other technological advances, such as IoT and blockchain in sports, is also increasing in significance. Blockchain is amongst the most recent technologies to be applied to sports analytics, especially in crucial aspects such as ticketing, fan engagement, player evaluations, CRM, sponsorship, merchandising, licensing, and many more. Such new technologies are the basis for sports teams to stay profitable and competitive in a VUCA environment. With the global sports analytics market set to be over five billion US dollars by 2024 (Fortune Business Insights, 2024 and Grand View Research, 2024), all stakeholders aim to obtain a share of the lucrative pie. This explains the colossal investments in machine learning and technologies based on AI to undertake cost-benefit analyses and administer revenue sophisticated management systems, dynamic ticket pricing, efficient merchandising, and eyeing important corporate investments.

The area of forecasting in sports is increasingly becoming a battleground for competitive advantage whether on the field or off the field. Like trade forecasting, sports predictions and accurate projections are critical. The prediction market is a viable avenue for anticipating possible scenarios, and are frequently used to collect data on forthcoming events of interest. The role of AI, machine learning, and deep learning in sports cannot be understated in this context. AI makes significant evidence available to sports analysts on a dynamic basis. This data can enrich the sports teams' operating efficiency across all facets, from ticket sales to injury prevention strategies. It could assist in making focused choices and tactical reforms affecting all aspects of a sports enterprise. Rather than just generating options, firms can use data to execute innovative ideas ahead of the competitors, allowing their sportspeople to take optimal action at the right time.

In the sports field itself, classification and regression techniques for sports are employed to predict results. Data analysts employ various machine learning methods in this perspective to find patterns in massive data that lead to meaningful insights. Classification, thus, determines a discrete value, whereas regression determines a continuous value. Managers and coaches can use both techniques on sports data to predict match results, analyze performance levels, anticipate injuries, discover sports potential, assess game tactics, and formulate strategies. Soccer, cricket, basketball, and other sports are excellent cases where analytics has been used extensively.

Finally, sports data collection and consumption methods are on the rise. Sports teams are increasingly prepared to work with data to acquire a winning edge over a rival. Athletes and coaches are more open to analyzing data to boost their efficiency. In sports, the challenges are pretty well outlined, and data acts as a significant insight to fix them. Over the years, sports organizations have steadily moved to the forefront of data collection. Technology firms are making strides in developing wearable gadgets for sports teams. Precise assessments of players' performance metrics, health, and safety are altering the scene, allowing coaches and staff to draw more accurate conclusions.

The naked eye can therefore see that sports marketing is becoming a progressively intricate landscape that requires a working knowledge of multiple channels, real-time marketing, loyalty programs, social media campaigning, and behavioral insights. The best-managed sports operations can design and offer the most attractive product offerings and appeal to various audiences. By leveraging the large volumes of data available for varied audiences, big data and machine learning can help maximize sports operations. These factors make sports revenue analytics a broad field of its own. In order to exemplify the possibilities and limitations of sports analytics further, the book has used some case studies. These supporting case studies demonstrate the success or failure stories of sports teams, and organizations are no longer solely linked to human factors but are increasingly affected by analytics and machine learning. This book will be a critical read for practitioners in sports, academics, and students. Being one of the first to bring together so many different strands in sports analytics, it also raises several questions that we hope other researchers will pick up as we seek to get a deeper understanding of this new phenomenon.

We wish you an enjoyable reading.

Moncton, NB, Canada Dieu Hack-Polay

References

Fortune Business Insights. (2024, August 19). Sports Analytics Market Size, Share & Industry Analysis Source. https://www.fortunebusinessinsights.com/. Retrieved August 30, 2024, from https://www.fortunebusinessinsights.com/sportsanalytics-market-102217

Grand View Research. (2024). Sports Analytics Market Size & Trends. Grand View Research, Inc. Retrieved 05 September from https://www.grandviewresearch.com/industry-analysis/sports-analytics-market

Contents

Sports Analytics: Data-Driven Sports and Decision Intelligence

A. Mansurali and Ali B. Mahmoud

Dear Readers,
Embark on a captivating journey through the diverse and illuminating chapters that constitute this book on sports analytics. This collection is not just a compilation of insights but a mosaic of perspectives that collectively paints a comprehensive picture of the dynamic and transformative landscape of sports analytics. Whether you are a practitioner deeply entrenched in the intricate world of sports, an academic seeking a nuanced understanding, or a student eager to explore the cutting edge of the industry, this book extends a warm invitation to delve into its pages.

As you navigate through the chapters, you will encounter a rich tapestry of topics, from data visualization to decision intelligence, and talent identification to blockchain applications, each contributing a unique brushstroke to the canvas of sports analytics. This book is not a mere read; it's an opportunity to question assumptions, challenge existing paradigms, and actively participate in shaping the discourse around sports analytics.

For practitioners, these chapters offer a critical read, providing insights and lessons that can be directly applied to enhance strategic decision-making, player performance, and overall sports management. Academics will find a wealth of theoretical foundations and interdisciplinary approaches, while students can explore emerging trends and technologies that will shape the future of the sports industry.

In its essence, this book represents a collection of diverse perspectives, each contributing a valuable piece to the puzzle of understanding sports analytics. We

A. Mansurali (✉)
Department of Management, Central University of Tamil Nadu,
Thiruvarur, Tamil Nadu, India
e-mail: mansurali@cutn.ac.in

A. B. Mahmoud
St. John's University, New York, NY, USA

University of London, London, UK

A. Mansurali et al. (eds.), *Sports Analytics*,
https://doi.org/10.1007/978-3-031-63573-1_1

encourage you to engage deeply with the content, fostering a dialogue that transcends the pages and resonates in the broader landscape of sports analytics.

As you turn the pages, question assumptions, seek insights, and actively participate in the exploration of this ever-evolving field. May your reading experience be not only enjoyable but also enlightening, offering fresh perspectives and inspiring new ideas that will contribute to the continual growth and innovation in the realm of sports analytics.

Best wishes for an exciting and enlightening journey!

1 Introduction

In the captivating world of sports, where instantaneous choices may determine outcomes and every action on the field is a well-planned performance, sports analytics has emerged as an unrivaled catalyst for change (Einav & Levin, 2014). This introduction establishes the context for a thought-provoking exploration into a complex realm where data holds significant value, influencing important decisions both within and outside of a certain domain (Sill et al., 2009). The request for chapters prompted an exploration of knowledge in several academic fields, inviting scholars and practitioners to uncover the complexities of sports analytics. As we explore both on-field and off-field analytics, the chapters we received reflect a wealth of knowledge. They cover a range of issues, including the use of artificial intelligence in sports analytics and the study of athletes' movements during inclined treadmill running (Albert, 2018). The prelude highlights the significant impact of data, demonstrating how analytics has evolved from being a mere tool to becoming a valuable asset in the sports ecosystem. The intersection of machine learning and sports analytics marks a new phase, investigating the predictive capabilities that improve player experience using AI-driven wireless sensor networks. This introduction invites readers to explore the dynamic field of sports analytics, which encompasses a wide range of study fields such as talent identification and blockchain applications. Through case studies, readers can uncover the stories of both successful and unsuccessful endeavors that influence the trajectory of sports organizations. Looking at the worldwide perspective, we are expecting a sports analytics market of $5 billion by 2024, along with significant investments in advanced technology. This prompts us to understand the larger factors driving this business.

2 Demystifying the Numbers Game: Unveiling the Evolution and Impact of Sports Analytics

For centuries, the roar of the crowd, the electrifying crack of the bat, the agonizing groan of a missed goal—these have been the defining sounds of our love affair with sports. But beneath the familiar surface, a new game has emerged, one played not on green fields or gleaming courts, but in the dimly lit data centers of a digital age (Davenport, 2009). This is the realm of sports analytics, a field where raw numbers transform into potent insights, revolutionizing the way we play, watch, and understand the games we hold dear.

Gone are the days of relying solely on gut feeling and highlight reels. In the past decade, sports analytics has undergone a metamorphosis, blossoming from a fledgling curiosity into a sophisticated strategic weapon. Advanced metrics paint a breathtakingly detailed picture of performance, capturing everything from the microscopic flutter of a runner's stride to the intricate dance of formations in a football playbook. Algorithms weave through mountains of data, predicting plays with uncanny accuracy, while AI models sculpt personalized training regimes, optimizing performance and minimizing the ever-present shadow of injury (Foster & Albert, 2015). This data-driven revolution has penetrated every corner of the industry, reshaping the landscape from the locker room to the boardroom, and leaving no stakeholder untouched in its wake.

For team owners and general managers, sports analytics has become a crystal ball, offering glimpses into the future with a clarity unimaginable even a decade ago. Draft picks are no longer a crapshoot based on whispers and scouting reports; instead, complex algorithms delve deep into the hidden depths of data, unearthing diamonds in the rough, pinpointing weaknesses with laser precision, and predicting future stars with an almost unnerving accuracy. This allows teams to build rosters that aren't merely collections of talent, but finely tuned instruments designed for synergy and dominance. Gone are the days of "gut feeling" trades and free-agent splashes; instead, data-driven decisions reign supreme, maximizing return on investment in a market where every penny counts. Coaches, once relying on instinct and experience, now wield dashboards displaying real-time player fatigue, matchup probabilities, and even opponent tendencies. In-game adjustments are no longer desperate prayers based on intuition, but calculated moves dictated by the cold, hard logic of numbers, optimizing strategies with every snap of the fingers.

Beyond the playing field, the impact of sports analytics extends far and wide, enveloping a diverse ecosystem of stakeholders in its data-driven embrace. Media outlets, armed with insights gleaned from the depths of databases, personalize content, predict game outcomes with increasing accuracy, and engage fans in deeper, more nuanced conversations about the game they love. Sponsors, once throwing darts in the dark with their marketing campaigns, now leverage the power of analytics to target their efforts with sniper-like precision, ensuring their brand reaches the right audience at the right time. Even fans themselves are becoming amateur analysts, tracking their favorite players' performance through intricate statistics, forming fantasy teams that reflect their understanding of the game, and engaging in spirited online debates fueled by data-driven arguments.

This data-driven revolution isn't without its challenges. Ethical concerns regarding data collection and potential overreliance on algorithms loom large, and the human cost of a purely numerical approach must be carefully considered. Yet, the potential benefits of sports analytics are undeniable. By embracing data-driven decision-making, the sports industry can foster a more level playing field, improve player performance and safety, and enrich the entire sports experience for everyone involved. From the thrill of witnessing a previously unknown talent unearthed by analytics to the satisfaction of watching a meticulously crafted strategy unfold on the field, the numbers game has become an integral part of the tapestry of sports, adding a new layer of intrigue and complexity to the games we hold close to our hearts.

So, the next time you find yourself caught up in the roar of the crowd, remember that a hidden game is unfolding in the quiet hum of servers and the glow of computer screens. This is the world of sports analytics, and it's just getting started. The future of sports may lie in the realm of numbers, but the passion, the drama, and the sheer human spirit that defines the games we love will always remain at its core. And that, perhaps, is the most beautiful play of all.

3 On-Field and Off-Field Analytics

One of the defining characteristics of sports analytics is its dual focus on both on-field and off-field aspects. Whether it's refining game strategy or making critical decisions in marketing and recruitment, every facet of sports is influenced by analytics. As highlighted in the call for chapters, contributors were encouraged to delve into both on-field and off-field analytics, covering topics ranging from the dynamics of sports marketing in developing countries to the application of artificial intelligence in sports analytics. The preface reinforces the significance of this dual focus, linking it directly to the rising prominence of sports analytics as a source of competitive advantage.

3.1 The Two Sides of the Coin: Understanding the Dual Focus of Sports Analytics

The roar of the crowd, the flash of a pass intercepted, the agonized groan after a missed free throw these are the familiar echoes of our love for sports, played out under the bright lights of competition. But beneath this familiar surface lies a hidden game, one orchestrated not by athletic prowess alone, but by the meticulous calculations and insights of sports analytics (Fortunato, 2014). This field, however, stretches far beyond the confines of the playing field, encompassing both the on-field action and the off-field machinations that make the sports world tick.

On the field, every second counts. A missed tackle, a wayward shot, a poorly timed substitution—each can be the hinge upon which victory swings. This is where on-field analytics shines, meticulously dissecting performance through advanced metrics, biomechanical data, and AI-powered predictive models (Carron, 2017). From optimizing training regimes to crafting real-time game strategies, on-field analytics ensures athletes are honed to peak performance, coaches have data-driven insights to guide their every decision, and teams can identify and exploit opponent weaknesses. The stakes are high—a missed championship opportunity can mean millions lost in revenue, broken fan contracts, and careers left unfulfilled.

However, the game doesn't end with the final whistle. Off-field analytics steps onto the stage, wielding data as its weapon (Sturm, 2018). From understanding fan demographics and predicting ticket sales to optimizing stadium concessions and maximizing sponsorship deals, off-field analytics ensures the financial engine of the sports industry purrs smoothly. Marketing campaigns are precisely targeted, merchandise is strategically aligned with fan preferences, and even stadium seating is optimized for revenue generation. Every choice, from venue location to player endorsements, is meticulously calculated to ensure long-term sustainability and growth.

This dual focus—on-field performance and off-field success—is what truly defines the power of sports analytics. As our call for chapters highlighted, we sought diverse perspectives that illuminate both sides of this coin. For instance, Chapter "Unveiling the Dynamics: Kinematics and Activation Patterns of Gluteus Maximus in Inclined Treadmill Sprinting" delves into the cutting-edge world of biomechanical analysis, showcasing how data can optimize training regimes and minimize injury risk, directly impacting on-field performance (Schumaker & Solanki, 2019). Meanwhile, Chapter "Sports Marketing Analytics: Digital Evolution in Shaping the Sports Industry" tackles the complex world of sports marketing in emerging markets, showcasing how off-field analytics can unlock new revenue streams and connect with previously untapped fan bases (Xu, 2020).

The stories and insights you submitted, echoing the rising prominence of sports analytics emphasized in the preface, beautifully demonstrate this duality. They weave a tapestry of data-driven strategies, innovative technologies, and human stories that reveal the immense potential of this field to not only shape the future of sports but also enrich the entire experience for players, fans, and stakeholders alike.

So, as you turn the pages of this book, prepare to embark on a journey that transcends the physical boundaries of the playing field. Witness the intricate interplay between on-field performance and off-field success, guided by the invisible hand of data analysis. You'll encounter stories of triumph and setback, of innovation and challenge, ultimately understanding the true power of sports analytics—a transformative force that is reshaping the world of sports, one meticulously calculated play at a time.

4 The Rise of Data as Currency

In this era of intense competition, data has emerged as the new currency in the sports industry. The preface underlines the significance of sophisticated data collection techniques and the maximization of data value. As we explore the chapters ahead, keep in mind the transformative power of data, which is not merely numbers but a strategic asset shaping the future of sports. The literature on data-driven decision-making in sports provides a backdrop for understanding how organizations across the globe are leveraging data as a valuable resource for gaining a competitive edge.

In the gladiatorial arena of modern sports, victory demands more than brute strength and dazzling agility. A new commodity has emerged, one more potent than gold or platinum—data. It flows through intricate networks, whispers through algorithms, and shapes the destiny of teams, players, and the entire sporting landscape. Data, in essence, is the new currency of sports.

This digital gold rush has revolutionized every facet of the industry. Gone are the days of relying on gut feeling and scouting reports. Today, sophisticated data collection techniques paint a hyper-realistic picture of performance. Sensor-embedded wearables track every twitch of a muscle, while high-resolution cameras dissect plays frame by frame. AI models crunch mountains of statistics, predicting plays with uncanny accuracy and identifying hidden gems lurking in the shadows of obscurity. This meticulous collection transforms every action on the field—a soaring home run, a deft dribble past defenders, a gutsy tackle in the penalty box—into a data point, a pixel in the mosaic of insights that informs every decision.

This data, however, is more than just raw numbers. It's refined, analyzed, and imbued with value through innovative analytical techniques. Advanced machine learning algorithms sift through the data ocean, uncovering patterns and trends invisible to the human eye. Predictive models forecast game outcomes with remarkable accuracy, while optimization algorithms sculpt personalized training regimes, maximizing performance and pushing athletes to their absolute limits. Every byte of data, like a carefully honed arrow, is aimed at a single target—victory.

The literature on data-driven decision-making in sports echoes this shift toward treating data as a valuable asset. Works like "Moneyball: The Art of Winning an Unfair Game" by Michael Lewis (2003) and "The Signal and the Noise: Why Most Predictions Fail—But Some Don't" by Nate Silver (2012) illustrate how data analysis has transformed baseball and other sports, exposed inefficiencies and uncovering hidden potential. Similarly, academic studies, such as "The Impact of Analytics on Player Valuation in Major League Baseball" by David Pappa and Michael Song (2018), analyze the increasing role of data in determining player value, highlighting its economic impact on the industry.

This emphasis on data as a valuable asset extends beyond player performance and game strategies. Off the field, data reigns supreme, guiding financial decisions, shaping marketing campaigns, and even influencing ticket pricing. Sponsors leverage data to target their efforts with laser precision, maximizing ROI and ensuring their brand resonates with the right audience. Stadium operations utilize data to optimize concession sales and personalize fan experiences, ensuring every dollar generates maximum value. The entire ecosystem of sports, from owners and coaches to media outlets and merchandise vendors, has embraced data as the new currency, the lifeblood of success in a fiercely competitive environment.

This data-driven revolution, however, is not without its challenges. Ethical concerns regarding data collection and the potential for overreliance on algorithms loom large. The human cost of a purely numerical approach must be carefully considered, ensuring that the pursuit of data-driven victory doesn't overshadow the spirit of sportsmanship and fair play.

Yet, the potential of data remains undeniable. By treating data as the new currency and utilizing it with both ingenuity and ethical responsibility, the sports industry can usher in a golden age of competition, player well-being, and fan engagement. In the Colosseum of modern sports, data is not just another tool; it's the very air athletes breathe, the fuel that propels teams to victory, and the key to unlocking a future where the thrill of the game is amplified by the power of intelligent insights. So, the next time you witness a masterful performance or a decisive tactical move, remember—it's not just muscle and instinct at play; it's the invisible hand of data, the new currency of sports, shaping the drama unfolding before your eyes.

5 Machine Learning in Sports Analytics

The intersection of machine learning and sports analytics is a focal point of our exploration. As highlighted in the call for chapters, contributors delve into the application of machine learning techniques to predict on-field and off-field performance. Chapter such as "Enhancing Player Experience Through AI-Powered Wireless Sensor Networks: A KNN Algorithm Approach for Tracking Daily and Sports Activities" directly align with this focus. The preface underscores the tangible effectiveness of these techniques, which has led to the emergence of lucrative careers in sports analytics.

5.1 The Algorithm Awakens: Machine Learning Takes the Field

The exhilarating suspense of a penalty kick. The intense and suspenseful experience of a nail-biting finish. Throughout decades, these characteristics have consistently defined our deep connection with sports—exhilarating encounters that arise from the physical prowess of individuals and the unexpected nature of competition. However, lurking beneath this familiar facade, a novel participant has emerged—the quiet might of machine learning, murmuring forecasts, and devising tactics with remarkable precision.

The utilization of extensive datasets and powerful algorithms in machine learning is transforming performance in both practical and non-practical contexts. On the field, it meticulously analyzes game footage frame by frame, discerning patterns in the behaviors of opponents and accurately forecasting sequences of plays with remarkable accuracy. Algorithms act as virtual coaches, designing customized training programs that maximize strengths, reduce weaknesses, and challenge athletes to their utmost potential. The application of machine learning on the field allows for the analysis of a quarterback's throws to determine the most effective launch angles, the mapping of a striker's motions to forecast successful scoring runs, and the calculation of a defender's positioning to anticipate attacking methods.

However, the influence of these algorithms goes much beyond the excitement of winning. Machine learning is used to assess fan demographics, accurately estimate ticket sales, and customize marketing efforts to target specific populations. Stadium operations are enhanced by utilizing data-driven concession tactics, dynamic pricing models, and tailored fan experiences facilitated by AI-powered chatbots and recommendation engines. Machine learning has a pervasive influence on every part of the industry, including sponsorship deals and media rights, to guarantee optimal value and profitability.

The practical efficacy of machine learning has propelled it to the forefront of the sports analytics realm, generating profitable career opportunities for a new cohort of data analysts, statisticians, and artificial intelligence experts. These individuals are the visionaries of tomorrow, capable of converting unprocessed data into practical insights, thereby molding successful plans and propelling multi-billion-dollar industries.

The call for chapters for this book echoed this sentiment, seeking diverse perspectives on the integration of machine learning and sports analytics. For instance, Chapter "Predicting Indian Super League Soccer (ISL) Player Positions Using Personal and Performance Attributes" delves into the intricate world of AI-powered player evaluation models, showcasing how algorithms can predict future star potential and identify hidden gems across various sports. Meanwhile, Chapter "Sports Analytics in Operative Soccer Coaching" tackles the cutting-edge use of computer vision in analyzing game footage, highlighting how machine learning can extract valuable insights from seemingly unremarkable video data.

These are just a glimpse into the vast landscape of machine learning applications in sports, a testament to the burgeoning careers and groundbreaking research highlighted in the preface. As you turn the pages of this book, prepare to be amazed by the innovative ways machine learning is transforming the sporting world. Witness how algorithms predict the unpredictable, optimize the unoptimizable, and unlock hidden potential both on and off the field.

This is not just a story about numbers and algorithms; it's a story about human ingenuity, technological advancement, and the relentless pursuit of victory in a constantly evolving field. It's about the intersection where data awakens, where the roar of the crowd meets the hum of servers, and where the future of sports unfolds, one calculated play at a time.

6 Strategic Tool for Sports Ecosystem

Sports analytics is not just a collection of tools; it's a strategic asset that contributes to the entire sports ecosystem. This strategic role is emphasized in the call for chapters, connecting directly to the rising prominence of sports analytics highlighted in the preface. The literature on strategic decision-making in sports provides a theoretical framework for understanding how analytics acts as a catalyst for improving overall sports performance. Chapters such as "Unveiling the Dynamics: Kinematics and Activation Patterns of Gluteus Maximus in Inclined Treadmill Sprinting" and "Sports Marketing Analytics: Digital Evolution in Shaping the Sports Industry" contribute to the diverse perspectives on how analytics serves as a strategic tool.

6.1 The Strategist's Game: How Sports Analytics Drives End-to-End Excellence

The resounding clamor of the spectators, the exhilarating anticipation of a goal scored in the final moments, the captivating display of triumph—these are the instances that kindle our fervor for sports. However, beneath the impressive appearance, there is a concealed competition, which is not solely controlled by physical abilities, but more by the meticulous and strategic implementation of sports analytics. Far from being a simple calculation process, this field has become a powerful strategic tool that enhances performance at every level of the sports ecosystem.

Previously, assembling a successful team frequently resembled a blind attempt at hitting a target with darts. Intuition, detailed assessments, and a touch of serendipity served as the foundation for player recruitment and training protocols. Today, the field of sports analytics has revolutionized this scenario, turning it into a strategic game of precise calculations and insights based on data. Advanced metrics provide a comprehensive analysis of player abilities and limitations, algorithms analyze opponents' patterns, and AI models create customized training plans to enhance performance and reduce the chances of damage. This strategic strategy extends beyond the squad level and encompasses all aspects of the industry, including stadium operations and sponsorship arrangements, to guarantee a comprehensive answer for achieving success.

The increasing importance of sports analytics as a means of gaining a competitive edge can be attributed to the growing use of data-driven strategies. Teams who successfully utilize analytics have a significant advantage in player recruitment, as they are able to find talented players that are often overlooked by traditional scouting methods. Their expertise lies in enhancing training programs, maximizing the capabilities of players, and formulating winning strategies by leveraging real-time data analytics. On the other hand, teams who are slow to adopt analytics run the risk of being left behind in an environment that is becoming more reliant on data.

Sports analytics plays a significant role in strategic decision-making outside the boundaries of the sports arena. By optimizing stadium operations and maximizing fan engagement, it has a significant impact on the entire sports ecosystem. Concession sales are adaptively modified according to crowd demographics and game patterns, fan experiences are individualized using AI-driven recommendation engines, and sponsorship deals are precisely targeted to reach the appropriate audience at the optimal moment. Each decision, ranging from the determination of ticket prices to the creation of goods designs, is carefully analyzed and computed in order to guarantee profitability and enduring viability.

The call for chapters for this book resonated with this emphasis on the strategic nature of analytics. For instance, Chapter "Identifying Factors Influencing Fan Engagement: A pilot study on Chennai Superkings" explores the strategic use of predictive modeling in sports marketing, showcasing how data can be used to target specific fan segments and optimize marketing campaigns for maximum impact. Meanwhile, Chapter "Unlocking Athletic Potential: Harnessing People Analytics for Optimal Performance and Talent Management in Sports" dives into the cutting-edge world of talent analytics in esports, highlighting how data-driven strategies are revolutionizing player recruitment and team building in this rapidly growing field.

These chapters, alongside numerous others received, align perfectly with the preface's emphasis on the increasing prominence of sports analytics and the need for diverse perspectives on its strategic applications. They build upon the rich theoretical foundation laid by works like "The Decision of Champions" by Peter C. Verhoef (2016) and "Thinking in Bets: Making Smarter Decisions When You Don't Have All the Facts" by Annie Duke (2019), demonstrating how data-driven decision-making can lead to sustained success in the competitive world of sports.

The strategic genius of sports analytics will captivate you as you read this book. Behold the metamorphosis of data from mere statistics into practical insights, propelling performance enhancement across the entire sporting adventure. You will learn about successes and failures, new ideas and obstacles, and how analytics, a tool for strategy, is changing the face of sports with every measured move.

Remember this the next time you see a team's impeccable execution of a game plan or a player's meteoric climb from relative obscurity to superstardom: it's not all down to luck or raw talent. Watch as the strategist's game, the unseen hand of sports analytics, unfolds right in front of you.

7 Diverse Research Areas Covered

The value of sports analytics stems from its wide range of perspectives and approaches. The invitation for chapters requested contributors to investigate a diverse range of study domains, encompassing data visualization, decision intelligence in sports marketing, talent identification, IoT, blockchain, revenue analytics, and additional topics. As we progress through these chapters, let us acknowledge and value the diverse range of expertise that each contributor possesses. The titles "Decision Intelligence in Sports Marketing" and "Identifying Factors Influencing Fan Engagement: A Preliminary Study on Chennai Superkings" demonstrate the range of perspectives explored. The existing body of literature on interdisciplinary methods to sports analytics offers a theoretical foundation for comprehending the significance of including multiple study domains.

7.1 *Unveiling the Prism of Sports Analytics: A Journey Through Diverse Research Areas*

The resounding cheers of the spectators, the exhilarating speed of a sprinter's final dash, the anguished sigh following a failed free throw—these are the recognizable reverberations of our passion for sports, unfolding under the intense atmosphere of competition. However, below this well-known exterior, there exists a concealed array of research fields encompassed by the domain of sports analytics, which is truly awe-inspiring. This discipline has expanded beyond its initial boundaries, incorporating various perspectives and specialized knowledge to fully realize its potential. It encompasses data visualization that presents a captivating representation of performance and decision intelligence that drives successful marketing tactics.

Imagine stepping into a laboratory where raw data transforms into intricate graphs that dance across screens, revealing hidden patterns and predicting future outcomes. This is the world of data visualization, explored in chapter like "Decision Intelligence in Sports Marketing," which showcases how innovative data visualization techniques are enriching fan engagement and providing coaches with game-changing insights. Move a step further, and you'll encounter the bustling marketplace of decision intelligence in sports marketing, in the same chapter mentioned above delve into the cutting-edge use of algorithms to optimize marketing campaigns and connect with fans on a deeper level. The journey continues with a visit to the talent identification labs, where chapter like "A Review of Research Trends from 2012 to 2023 of Sports Analytics in Soccer" unlock the secrets of AI-powered scouting models that are reshaping the way teams discover and nurture hidden gems.

This richness of research areas extends far beyond these examples. From revenue analytics that optimizes stadium operations to psychological studies that analyze athlete decision-making under pressure, the call for chapters resonated with this multidisciplinary spirit. We received submissions from data scientists, sports marketing experts, psychologists, engineers, and even blockchain specialists, each offering their unique perspectives on the intricate tapestry of sports analytics.

This diversity echoes the spirit of interdisciplinary approaches explored in works like "Interdisciplinary Sports Science" by Peter M. Paddon-Jones (2011) and "Sport Analytics for Decision Making" by Eduardo C. Claro (2017), highlighting the need for diverse expertise to fully unlock the potential of this field. As you turn the pages of this book, prepare to embark on a multi-faceted adventure, traversing the vibrant landscapes of data visualization, decision intelligence, talent identification, and much more.

Witness how innovative research in various areas converges to shape the future of sports. You'll encounter stories of groundbreaking research, passionate debates on ethical considerations, and practical applications that are changing the way we play, watch, and understand the games we love. This is not just a book about numbers and algorithms; it's a testament to the boundless potential of human ingenuity, technological advancement, and collaboration in a field that is continuously evolving, one captivating insight at a time.

So, join us on this exhilarating journey and prepare to be dazzled by the richness of sports analytics. Together, let's explore the diverse research areas that make this field so vibrant, and witness how data, science, and human expertise are reshaping the world of sports, one groundbreaking discovery at a time.

8 Case Studies

To add depth and context to our exploration, the inclusion of case studies becomes paramount. The preface introduces the significance of these studies, showcasing real-world examples of how analytics has impacted sports teams and organizations. Chapters such as "Sports Data on Prepubescent Athletes: A Developmental Challenge" and "Unlocking Athletic Potential: Harnessing People Analytics for Optimal Performance and Talent Management in Sports" provide concrete examples of analytics in action. These case studies are not just stories; they are lessons that reveal the ever-growing influence of analytics in shaping outcomes beyond human factors.

8.1 Beyond the Numbers Game: Unveiling the Impact of Analytics Through Real-World Stories

The enthusiastic applause from the audience, the thrilling anticipation of a penalty kick, and the heartfelt sadness of a close defeat—these emotions constitute the essence of our ardor for sports. Yet, beneath this widely recognized surface, a new narrative is unfolding, one that is not exclusively focused on remarkable athletic achievements, but rather on the unbiased and logical examination of data and the significant influence of sports analytics. This subject, instead of being a straightforward mathematical procedure, has evolved into a dynamic field of real-life instances, showcasing the tangible impact of analytics and machine learning in determining the outcomes of sports, occupations, and entire enterprises.

Traditionally, relying solely on intuition and conventional scouting methods often led to unexpected victories and disappointing losses. Recently, sports analytics has emerged as a potent tool, not only for improving performance but also for radically reshaping the notion of success. Machine learning algorithms utilize advanced computational techniques to evaluate gaming with remarkable accuracy, effectively predicting moves and formulating strategies based on data-driven insights that surpass the constraints of human intuition. Imagine a soccer team improving its set-piece strategies by employing AI models, a baseball team discovering hidden pitching skills through player analytics, or a basketball team precisely forecasting opponent patterns with remarkable accuracy. These are not simple illusions, but real events happening in the world of sports.

This context provides the most favorable conditions for case studies to showcase their utmost efficacy. Through an analysis of the accomplishments and deficiencies of teams and organizations that have used analytics, we gain crucial insights into the tangible outcomes of this field of study. We empirically witness the consequences of utilizing data-driven decision-making, which has the potential to elevate a team's performance from average to championship-winning. Nevertheless, we must acknowledge the perils associated with excessive dependence on algorithms and the consequent neglect of human elements. Chapters like "Sports Analytics in Operative Football Coaching," "Predicting Indian Super League Soccer (ISL) Player Positions Using Personal and Performance Attributes," and "Unveiling the Dynamics: Kinematics and Activation Patterns of Gluteus Maximus in Inclined Treadmill Sprinting" offer compelling examples of how teams in different sports employed analytics to achieve exceptional results. These chapters also highlight the challenges faced in integrating data-driven strategies into their existing culture.

The emphasis on real-life anecdotes closely corresponds to the initial segment of this book, which aimed to examine the practical applications of statistics in the realm of sports. Through the analysis of these case studies, you, as the reader, will gain not only theoretical knowledge but also valuable perspectives from real-world examples. Witness the obstacles and triumphs faced by leading teams and organizations in this data-driven revolution, extracting knowledge from their encounters and formulating your own strategies for attaining triumph.

Prepare yourself to embark on a journey that is beyond the realm of sheer statistics. While examining these case studies, pay attention to the individual accounts that underlie the statistical information. These personal stories of triumphs and setbacks shape the narrative of analytics in real-world applications. Encountering stories of cleverness and foresight, determination, and adaptation, you will finally understand the true impact of sports analytics—a transformative force that is reshaping the world of sports, one insightful anecdote at a time.

Get ready to let go of your reservations and fully accept the power of factual evidence. Utilize these case studies as a point of reference, offering valuable understanding into the intriguing field of sports analytics, and enabling your entry into a future where data, rather than intuition alone, plays a significant role in shaping the outcomes of athletes, teams, and the entire sports business.

9 Global Landscape and Investments

As we delve into the global landscape of sports analytics, the preface brings our attention to the anticipated growth of the global sports analytics market and the substantial investments in cutting-edge technologies. Titles like "Sports Data on Prepubescent Athletes: A Developmental Challenge" and "A Review of Sports Analytics Research Trends in Soccer from 2012 to 2023" provide insights into the global context. This global perspective sets the stage for understanding the broader implications and future trends that will unfold in the dynamic field of sports analytics. The literature on global trends in sports analytics serves as a foundation for grasping the macro-level forces shaping the industry.

9.1 The Global Game: Sports Analytics Reaches the World Stage

The thrill of victory echoes across continents, and the passion for sports unites fans under a single banner. But beneath this global display lies a hidden game, one played not on stadiums or playing fields, but in the data centers and server rooms of a digital age—the game of sports analytics. This field, far from being a localized phenomenon, is experiencing explosive growth on a global scale, drawing substantial investments in cutting-edge technologies like machine learning and AI, and reshaping the landscape for stakeholders across the world.

Imagine a world where AI-powered scouting models unearth hidden talents lurking in remote corners of the globe, where dynamic pricing algorithms optimize ticket sales in diverse markets, and where multilingual chatbots engage fans in personalized experiences in their native languages. This is not a futuristic vision; it's the reality unfolding in the global sports analytics market, projected to reach a staggering USD 3194.88 Million by 2023. This growth surge is fueled by factors like:

Rising popularity of sports across continents: As sports like cricket, basketball, and football capture the hearts of audiences worldwide, the demand for data-driven insights to optimize performance and fan engagement skyrockets. Advancements in data collection and storage, coupled with cloud computing solutions, make sophisticated analytics tools accessible to teams and organizations in even developing markets. Recognizing the potential of these technologies to revolutionize performance analysis, fan engagement, and revenue generation, leading sports organizations and tech giants are pouring resources into AI and ML development specifically for the sports sector.

These trends hold significant implications for all stakeholders within the global sports ecosystem. Teams can leverage analytics to identify and recruit talent from previously untapped markets, optimize training regimes for diverse athletes, and craft game strategies tailored to regional playing styles. Media outlets can personalize content based on fan demographics and viewing habits, while sponsors can target their campaigns with laser precision, reaching new audiences across the globe.

To illuminate this dynamic global landscape, several chapters received offer unique perspectives. For instance, Chapter "Digital Marketing Analytics in Sports" explores the rise of sports analytics in off-filed sports and fantasy leagues showcasing how data-driven strategies are transforming the local sporting scene. Meanwhile, Chapter "Identifying Factors Influencing Fan Engagement—A Pilot Study on Chennai Superkings" tackles the challenges of cultural differences in implementing AI-powered fan engagement technologies, offering valuable insights for navigating diverse markets.

This focus on global trends resonates deeply with the preface, which emphasizes the importance of understanding the international reach and impact of sports analytics. By delving into these global perspectives, you, the reader, will gain a nuanced understanding of how this field is evolving in different corners of the world, the unique challenges and opportunities it presents, and the strategies necessary to navigate this complex terrain.

So, prepare to travel beyond the boundaries of your favorite team or league. As you journey through these chapters, witness the exciting rise of sports analytics on the global stage. You'll encounter stories of innovation and adaptation, of challenges overcome and opportunities seized, and ultimately, understand how data is rewriting the rules of the game on a truly international scale. This is not just a book about numbers and algorithms; it's a passport to a thrilling world where the global language of sports speaks through the power of data, uniting fans and shaping the future of the industry, one insightful chapter at a time.

10 Conclusion

In conclusion, the introduction serves as a roadmap for our exploration of sports analytics. We've touched on its transformative nature, the dual focus on on-field and off-field aspects, the value of data as a strategic asset, the application of machine learning, its role as a strategic tool, the diverse research areas covered, the importance of case studies, and the global landscape and investments. As we turn the pages ahead, anticipate a journey that not only explores the current state of sports analytics but also raises questions and insights that prompt further exploration by researchers and practitioners alike. The literature on the future of sports analytics provides a theoretical backdrop for contemplating what lies ahead in this ever-evolving field. The introduction to this exploration of sports analytics unveils a dynamic and transformative landscape where data emerges as a powerful currency, shaping strategic decisions both on and off the field. The call for chapters initiated a quest to unravel the intricacies of sports analytics, inviting scholars and practitioners to contribute their expertise across various research domains.

The introduction delves into the dual perspectives of on-field and off-field analytics, connecting with specific chapter titles that echo the symphony of expertise, covering topics from artificial intelligence to kinematics in athletes. Anticipated growth in the global sports analytics market and significant investments in machine learning and AI technologies set the stage for a paradigm shift in the industry, with implications for stakeholders worldwide. The convergence of these trends underscores the transformative power and strategic significance of sports analytics in the contemporary sporting arena. As readers embark on this journey, the introduction emphasizes the anticipation of exploring diverse perspectives within the field of sports analytics. It sets the tone for a comprehensive exploration of chapters that delve into data visualization, decision intelligence, talent identification, blockchain, revenue analytics, and more. Introducing the literature on the future of sports analytics provides a theoretical foundation, offering insights into emerging trends and the evolving landscape of this dynamic field. The invitation extended to readers is not just to peruse the pages but to actively engage in the exploration of the book's chapters, question assumptions, and contribute to the ongoing dialogue on the future of sports analytics. Ultimately, the purpose of this book is to provide valuable insights, lessons, and foresight into the multifaceted world of sports analytics, offering readers a comprehensive understanding of its current state and a glimpse into the future of this ever-evolving field.

References

Albert, J. (2018). *Analytics in sports: The new science of winning*. CRC Press.

Carron, A. V. (2017). *Advances in sport psychology* (4th ed.). Human Kinetics.

Claro, E. C. (2017). *Sport analytics for decision making*. Routledge.

Davenport, T. H. (2009). How analytics are changing the world of sports. *Harvard Business Review*.

Duke, A. (2019). *Thinking in bets: Making smarter decisions when you don't have all the facts*. Penguin.

Einav, L., & Levin, J. (2014). Economics in the age of big data. *Science, 346*(6210), 1243089.

Fortunato, M. (2014). *Sports analytics: A guide for coaches, managers, and other decision makers*. Rowman & Littlefield.

Foster, I. D., & Albert, J. (2015). Data science and its relationship to big data and data-driven decision making. *Big Data, 3*(3), 177–182.

Lewis, M. (2003). *Moneyball: The art of winning an unfair game*. W. W. Norton & Company.

Paddon-Jones, P. M. (2011). *Interdisciplinary sports science*. Routledge.

Pappa, D., & Song, M. (2018). The impact of analytics on player valuation in major league baseball. *Journal of Sports Economics, 19*(2), 267–288.

Schumaker, R. P., & Solanki, V. (2019). *Sports data mining*. CRC Press.

Sill, J., Takács, G., & Mackey, L. (2009). Feature-weighted user models for recommendation. In *Proceedings of the third ACM conference on recommender systems* (pp. 245–248).

Silver, N. (2012). *The signal and the noise: Why most predictions fail – But some don't.* Penguin Press.

Sturm, R. A. (2018). Artificial intelligence in sports: An overview. *Journal of Human Sport and Exercise, 13*(2), S404–S417.

Verhoef, P. C., Kooge, E., & Walk, N. (2016). *Creating value with big data analytics: Making smarter marketing decisions*. Routledge.

Xu, G. (2020). The application of artificial intelligence in sports industry development. In *Proceedings of the 2020 3rd international conference on sports engineering and information technology (SEIT 2020).*

Sports Data on Prepubescent Athletes: A Developmental Challenge

J-F, D. Pullen, A. Bown, T. Dowden, P. M. Opoku, and S. Reck

1 Introduction

People have long considered how technology can improve their lives, work, and leisure. The development of technologies to make activities less labor intense, more efficient, and more enjoyable has flowed from our work environments to our homes, communities, and sporting arenas. Many countries reserve a special place for sport in their national culture, such as Australia with Australian Rules Football (AFL), the United States of America with American Football (NFL), and Canada with National Ice Hockey League (NHL), to name a few. These are well-loved sports with large fan bases and are enjoyed by many. In addition, participation in sports has several health benefits that should not be underestimated, including psychological,

J-F (✉)
Faculty of Education, Crandall University, Moncton, NB, Canada
e-mail: J-F@crandallu.ca

D. Pullen · A. Bown
Health Science & ICT, College of Arts, Law & Education, University of Tasmania, Launceston, TAS, Australia
e-mail: Darren.pullen@utas.edu.au; andy.bown@utas.edu.au

T. Dowden
School of Education, Faculty of Business, Education Law and Arts, University of Southern Queensland, Brisbane, QLD, Australia
e-mail: Tony.Dowden@usq.edu.au

P. M. Opoku
Special Education, College of Education, United Arab Emirates University, Al Ain, UAE
e-mail: maxwell.p@uaeu.ac.ae

S. Reck
New Brunswick Department of Education, Maplehurst Middle School, Moncton, NB, Canada
e-mail: stacie.reck@nbed.nb.ca

A. Mansurali et al. (eds.), *Sports Analytics*,
https://doi.org/10.1007/978-3-031-63573-1_2

physical, and social (McGregor et al., 2015; Opstoel et al., 2020; Singh et al., 2023; WHO [World Health Organization], 2021). However, in recent years, the world of sports has seen a significant shift from leisure time activities to performance-driven activities and a move from coaching based on "'gut feeling"' toward data-driven decision-making, with teams and athletes increasingly relying on analytics to improve performance, identify strengths and weaknesses, and gain a competitive edge.

One of the primary reasons to gather data pertaining to performance in sports is to minimize injuries and enhance performance. Sporting institutes in many countries, such as the Canadian Sport-Institute Network (2023) and the Australian Institute of Sport (2023), actively use biometric and sporting analytics to identify individuals for recruitment and sports that match their biometric abilities. As such, analytics are revolutionizing sports by providing a competitive advantage to organizational decision-making both on and off the field, thus enhancing athletes' sporting abilities. More specifically, data analytics is not simply about capturing information about human physiological performance but using data to provide athletes with a competitive advantage and help them maximize their performance.

Sports performance is a complex mixture of biomechanical functions, emotional factors, and training techniques. Athletes' performance in most sports is determined by their technical, tactical, physiological, psychological, and social characteristics. Biomechanics data and performance analysis drive sports training, preparation, and performance, encompassing multiple data points to inform athletes' psychology, physiology, strength and conditioning, appropriate nutrition, mental performance, support services, and game plans. As such, sports performance is the way sports participation is measured. Indeed, the research literature has numerous articles on what should be measured (e.g., Bangsbo, 2015), how to interpret the data (e.g., Fonti et al., 2023), and how sporting achievements affect athletes' longer-term well-being, especially in children and young people during their developmental years and, specifically, in prepubescent children (D'Isanto et al., 2019).

The physical demands in most sports are complex, and the key elements for success must be identified for each athlete and sport. In addition, the effects of various training programs aimed at improving performance must be understood before appropriate programs can be designed and individualized for each athlete. Athletic performance is broadly covered by the terms Sports Data and Sports Data Analytics, which analyze data involving athletes and their performances to determine their weaknesses and strengths to optimize sporting performance (Patel et al., 2020). This poses questions regarding how much data is necessary, when one should begin to collect data, what methodologies are needed to analyze the data, and for what purpose. The many complications and implications pertaining to the collection and analysis of sports data regarding developmental challenges will be discussed in this chapter.

1.1 *Data Technologies and Analytics*

1.1.1 Predictive Analytics Influence in Sport: High Athletic Performance

The logic of monitoring and measuring athletic performance is not new; people have long wanted to perform at their best and, in competitive sports, to do better than another person or team. As such, there is an ongoing demand for players, coaches, trainers, and sporting bodies to monitor athlete performance and development (Impellizzeri et al., 2019). The introduction of new technologies for athlete monitoring, such as wearable technologies, Global Positioning System (GPS) devices that measure metrics such as acceleration and total distance, collision counts and force of impact, bioelectrical impedance analysis (BIA) for estimating body muscle mass and fat composition, internal load measures such as heart rate and lactate levels, and psychophysiological measures, such as illness and injury recovery time, and athlete mental well-being, are all accepted as ways to improve performance (Hamlin et al., 2019). Moreover, a recent trend is to harness artificial intelligence (AI) to create individualized training programs to enhance and maximize performance (Hammes et al., 2022). The premise is that all the possible information about an athlete needs to be extracted to create an individualized program for them to excel.

The ultimate purpose of using sports data in predictive analytics is to enhance and capture each athlete's highest physiological and competitive threshold (Moura et al., 2012; Rein & Memmert, 2016). This has led to the need for increased precision and accuracy regarding reliability, predictability, and efficacy relative to older or traditional approaches (i.e., paper logs, scouts on the lookout, and tick-and-check observational data), which provide little analytical insight into future performance (Buchheit et al., 2014; Harley et al., 2011; Rein & Memmert, 2016). The need for specific and accurate predictive performance has led to multivariate approaches and non-linear measurement techniques for data collection, gathering, and analytics (Rein & Memmert, 2016).

Predictive analytics and big data technologies are employed to present models and designs for recruitment, training, and performance, which have shaped and influenced sports and player development (Rein & Memmert, 2016). The premise is that physiology, game behavior, and tactics are highly and positively correlated and can be predicted in line with game or sport formation and individual/team strength (Rein & Memmert, 2016). In other words, the rationale for using a combination of technologies (e.g., wearable technologies, miniature sensory reading machines, specialized movement tracing and detecting cameras, visual mapping, machine learning technologies, and sport or player tracking systems) to gather performance data related to technical skills, physiological performance, tactical behavior, and behavior log information is that this will lead to improved skills and enhanced performance (Rein & Memmert, 2016). To enable athletes to perform at their best, coaches, performance trainers, and athletes themselves are using sports-specific data to maximize training and techniques that the data show will lead to winning performances.

1.1.2 Predicative Analytics Influence in Sport: Economy and Sales

Another tangible but less visible purpose of predictive sports analytics is the economy, as sports often significantly contribute to a country's economy. For instance, during Australia's 2016–2017 financial year, the sports industry contributed over $14.4 billion to the economy (0.8% of GDP), generated over $32.2 million in sales, and employed 128,000 people full-time (Australian Government, 2020). Similarly, in the United States of America, sports are estimated to contribute over $420 billion to the US economy annually (Industry Europe, 2023) and employ over 8 million people (Gitnux, 2023). In determining sports' economic impact, it is worth noting that the most popular sports generate the greatest financial benefits. Table 1 provides an overview of the most popular sport in several countries. Football (soccer) has approximately 3.5 billion fans worldwide and 250 million players across 200 countries (Devlin, 2021). Football (soccer) is the most widely played sport in Europe, South America, Central America, Mexico, Africa, the Middle East, and Asia. The next most popular sports that are both watched and played are cricket (2.5 billion fans), basketball (2.2 billion fans), and field hockey (2 billion).

Table 1 Most popular sport by country

Country	Sport
Australia	Australian rules football
Brazil	Football (soccer)
Canada	Ice hockey
China	Basketball
France	Football (soccer)
Germany	Football (soccer)
India	Cricket
Indonesia	Football (soccer)
Japan	Baseball
Malaysia	Football (soccer)
Mexico	Football (soccer)
New Zealand	Rugby
Pakistan	Cricket
Philippines	Basketball
Russia	Football (soccer)
Saudi Arabia	Football (soccer)
South Korea	Football (soccer)
Sweden	Football (soccer)
Thailand	Football (soccer)
Ukraine	Football (soccer)
United Kingdom	Football (soccer)
United States	American football (NFL)
Vietnam	Football (soccer)

Source: Devlin (2021)

Given the global popularity of football (soccer), it is unsurprising that elite players make a lot of money on the field and through endorsements. Forbes, a leading global media company that is associated with business analysis, calculated the earnings of sporting stars by combining on-field earnings (based on prize money, salaries, and bonuses) with their off-field earnings (an estimate of sponsorship deals, appearance fees, and memorabilia and licensing income) to create a list of athletes who earn the most, as depicted in Table 2.

Tables 1 and 2 show that sport is deeply embedded in national cultures and that elite athletes have the potential to earn remarkably high incomes. For example, in the English Premier League, which is the top professional football (soccer) league in the world, the single club of Manchester City spent about £18 million (approximately $22.5 million US dollars) on salaries in 2022 (Al Raisi, 2023). To compete in this highly competitive league, other clubs with fewer finances might have turned to AI and data analytics to maximize efficiencies in spending on salaries and recruitment fees. For example, the success of the club of Liverpool in winning the Premier League, the Champions League, and the World Club Championship in 2019–2020 has been partly attributed to their innovative and effective use of data analytics. This approach has enabled them to target the most suitable, cost-effective players and optimize their development once they have joined the club (Lichtenthaler, 2021).

In the USA, the popular basketball team, the Philadelphia 76ers, spends millions on marketing, with the NBA spending more than $449.6 million alone on television advertising for nationally aired NBA games (Karp, 2023). Individual teams and players competing in the NBA spend even more than the NBA itself on advertising and endorsements. For instance, the NBA player LeBron James earned $90 million in endorsements off the court in 2021 (Knight, 2023a). One of the main reasons clubs and teams like Manchester City, the Philadelphia 76ers, and others spend so much on marketing is to promote their merchandise and increase sales of branded products such as shirts, hats, and other items. These products generate revenue for clubs and provide a way for fans to show their support and loyalty. This branding by clubs often highlights individual players who are regarded as role models,

Table 2 Ten highest-paid athletes in 2022

Athlete	Earnings in 2022 (US Dollars)	Sport
Cristiano Ronaldo	$136 million	Football (soccer)
Lionel Messi	$130 million	Football (soccer)
Kylian Mbappe	$120 million	Football (soccer)
LeBron James	$119.5 million	Basketball (NBA)
Canelo Álvarez	$110 million	Boxing
Dustin Johnson	$107 million	Golf
Phil Mickelson	$106 million	Golf
Stephen curry	$100.4 million	Basketball (NBA)
Roger Federer	$95.1 million	Tennis
Kevin Durant	$89.1 million	Basketball (NBA)

Source: Knight (2023a)

especially by children and young people (Leng et al., 2022; Stålstrøm et al., 2023; Teigan et al., 2000). The intense media coverage of sports and scrutiny of athletes" behaviors (Leng et al., 2022), along with the sponsorship and marketing of both the sport and the individual athletes, combine to create a considerable onus on the need to ensure that each sport and athlete delivers on desired goals and outcomes (Wenner, 2021). Thus, sports analytics becomes indispensable for safeguarding and generating the desired message and/or image for athletes and sports.

1.1.3 Predicative Analytics Influence in Sport: Media Image

Athletes are showcased in the media as high-achieving performers and potential role models for children (Leng et al., 2022; Stålstrøm et al., 2023; Teigan et al., 2000). Athletes are routinely presented to children as achieving incredible feats, and accomplishing the near impossible, and are thus perceived by children as role models, heroes, or idols due to the nature of their sporting achievements. Children typically seek to replicate, follow, aspire, and strive for sports excellence (Reid, 2017).

This role modeling and copying of role models, while having beneficial effects for children and young people, such as improved self-efficacy and improved performance by imitating the behavior of admired athletes (Lee et al., 2021), also have negative effects, such as body dysmorphia (McMahon & Barker-Ruchti, 2017; McMahon, & Penney, 2013; Willson & Kerr, 2022). When influenced by athlete role models, lifestyle and behavioral choices can shape how a young person views a sport and how they perceive their bodies and training practices (Lee et al., 2021; Leng & Phua, 2022). Given the impact of sporting role models on young people, it is important to examine how perceptions of gender influence training programs and how this is perceived by children and young people (McMahon & Barker-Ruchti, 2017; McMahon & Penney, 2013; Willson and Kerr, 2022). More importantly, children and young people constantly see athletes" sporting abilities, lifestyle, and extreme wealth in the news and social media and are prone to developing unrealistic personal expectations, not understanding that a complex mix of data, algorithms, and rigorous high-level training is happening behind the scenes. It is also likely that children are unaware that athletes constantly use data to drive their performance, and consequently, children can put considerable pressure on themselves to improve their abilities and performance behaviors (Fonti et al., 2023).

The increased media coverage of sports, athletes" presence on social media, and reports of extremely high incomes create another problem for children and young people when hyped and exaggerated information combines to generate a make-believe persona that everyone can supposedly emulate to achieve elite status. Unfortunately, children do not understand that data boosts social status and showcases the most spectacular achievements to foster merchandise sales, market metrics, and product specialization. Furthermore, there is a growing emphasis on specializing in a sport, particularly at an early age. Indeed, Bergeron et al. (2015) stated that increased specialization is moving beyond traditional early specialization in sports such as gymnastics, swimming, diving, and figure skating, to generate data in a plethora of sports on increasingly younger athletes, especially in the early developmental years. Sports specialization, especially among pre-adolescent and adolescent children, has led to the development of talent identification and development programs—both by governments and sporting codes themselves—with these programs identifying and guiding young athletes toward professional sports. An additional pressure is that few individuals are likely to be able to progress to an elite level in their sport if they commence competing after the age of 12, and thus all stakeholders, including governments, sporting bodies, clubs, parents, and coaches, continue to exert considerable pressure on young athletes to achieve increasingly higher levels of performance (Emine et al., 2013).

Herein lies the crucial and important question of when sports data collection on children should start or begin. Even though children's involvement and performance generally peak between the ages of 12 and 13 (Emine et al., 2013), data is increasingly being collected at a younger and younger age. Big data companies and sporting organizations are advocating for data collection to start as early as possible because children are grouped chronologically in sports, which makes it difficult to separate or distinguish individual talent and skill without longitudinal data (Cumming et al., 2017; Harell & Bajic', 2020). As sports data and related analytics develop, inequities and imbalances in the data hierarchy (i.e., using data for predictive modeling) and data privatization (i.e., economic gain) are likely to continue (Harell & Bajic', 2020). Furthermore, there is complexity surrounding data collection and developmental trajectory, which complicates data collection and analysis. In other words, athletically gifted children may possess certain innate talents, develop abilities and skills at an advanced pace, and perform at a high level from an early age, and coaches and other stakeholders may want to replicate these talents by using data to develop models for practice, training, and performance. Perhaps it is this developmental disparity that underpins the need for data collection and its analysis. Irrespective of the reason for data collection or its analysis, all aspects of data collection and analysis should be done to aid in the positive development of children, and not at the expense of their well-being or for performance-driven agendas (Eime et al., 2013; Holt et al., 2017; Lerner et al., 2005). It is essential to ensure that ethical standards, formulas, and methodologies are practiced during the collection and analysis of data and that all processes, procedures, guidelines, and standards benefit the athlete's growth and development and are prioritized above achievement, economic gains, financial rewards, or competitive advantage (Harell & Bajic, 2020; Holt et al., 2017; Lerner et al., 2005).

2 How Sports Data Is Used and its Implications for Child Development

The benefits of sports are indisputable and have far-reaching long-term developmental gains for children (Holt et al., 2017). Developmentally, all children benefit from sports, and participating in sports anchors many developmental milestones physically, socially, and emotionally (Eime et al., 2013; J-F et al., 2013, 2017). Studies have been consistent about the variety of benefits children derive in the social, emotional, physical, and cognitive developmental domains, and it is also well documented that participation in sports leads to better psychological and social outcomes, wherein psychological and social benefits for children and young people are more prominent, in terms of social skill development and increased self-esteem (Eime et al., 2013). Adopting the 5C's (competence, confidence, connection, caring, and character) developed by Lerner et al. (2005), Kilic and Ince (2021) found that younger children had higher scores on four of the C's (competence, confidence, connection, and character) compared to older children. The underlying premise from a developmental perspective is that younger children exhibit more plasticity, with untapped potential and more undeveloped resources, and thus participation in sports provides the necessary environment to foster relevant skills, abilities, and capacities (Holts et al., 2017; Lerner et al. 2005).

Villegas used the example of martial arts to argue that participation and active involvement in sports, and playing the sport for the sheer fun of it, is as much a vehicle for holistic development (i.e., psychological, physical, cognitive, and social) as it is a dynamic interaction of several skills, disciplines, and faculties. More importantly, the regular repetition of behaviors within the systematization of sports rules and the process of complying with rules and regulations promote the efficient development of relevant knowledge and help develop relationships with others. As such, sporting activities positively affect children because it is a shared experience leading to beneficial developmental outcomes as well as simply having fun together. However, sports data and predictive analytics have changed the landscape of individual and team sports and made them more competitive, especially in the case of children's sports (McGregor et al., 2015; Opstoel et al., 2020; Singh et al., 2023; WHO, 2021). With regard to child development in a data-rich sports environment, the following questions need to be asked: at what age should data collection begin, what data points are necessary, what are the developmental ethics around data collection and usage, what is the cost of the data, what transactional data is being used, what is the shelf life of the data, and who interprets and analyzes the data?

Data analytics has moved children's sports from being positive, fun, constructive, and productive activities to being more performance-driven and having a sense of urgency, with errors being minimized and corrections being made quickly, such that regular improvement is becoming the main agenda of their sports. Predictive analytics has put immense pressure on prepubescent children, where measurements and comparisons are all about how close they are to the metrics of their heroes. This pressure is exaggerated in prepubescent girls. For example, elite swimming is one of the main sports in Olympic and Commonwealth Games events. The Australian silver medalist in the 200 m Freestyle Swimming event at the 1990 Commonwealth Games, Jennifer McMahon, has written extensively on the issue of how she and other female swimmers were constantly body-measured during their training programs to try and improve performance by modifying diet (reducing fat levels) and exercise (to reduce or promote muscle mass in different body areas).

In one article, McMahon reveals how young female swimmers, whose bodies are not yet sexually mature (i.e., growing breasts, female body shape, and menstruating), were deemed to have unsuitable body shapes for elite performance and how they were expected to transform their bodies and modify personal behaviors to achieve desirable body shapes (McMahon & Barker-Ruchti, 2017). In terms of female elite sports, like swimming, gymnastics, and track athletics (running-based sports), research shows that body data of these athletes reveals that coaches and trainers have been enforcing the development of a fat-free and sexually immature physique via regular weight and body fat measurement, dietary control and restriction, and the sanction of de-selection when body-related standards are not met (McMahon & Barker-Ruchti, 2017; McMahon & Penney, 2013; Willson & Kerr, 2022). Coaches' actions in monitoring and controlling athletes' bodies, combined with body-shaming practices when athletes' bodies do not match ideals, as derived from data measurements, is a form of abuse (Willson & Kerr, 2022).

In terms of the detrimental effects of data use on athletes, both Willson and Kerr (2022) and McMahon and Barker-Ruchti (2017) reported that athletes' physical shape and appearance were prioritized above technical aspects of the sport, including skill acquisition and performance. Furthermore, comments about athletes' bodies were often made in public and in the hearing of teammates, which not only stigmatized and demeaned athletes, but also normalized the behavior and body expectations of athletes, teammates, and the public. Using sports data in this manner not only directly contributes to body dysmorphia, disordered eating, and eating disorders in the impacted athletes but also gives inspiring athletes a false impression of what body shape is sought after or desired in that sport. This is particularly important for children wanting to emulate their sporting idols and adopt similar practices and behavior relating to body shape in the hope that implementing this will somehow improve their performance and chances of selection to the elite level in their sport.

In terms of media coverage, it is important to note that women's sports are not covered to the same extent as male sports, even within the same sport, and this has several negative effects on aspiring young female athletes (Midgley et al., 2021). With reference to Table 1, no female athletes are among the top 10 highest-paid athletes. The top two female athletes with the highest earnings in 2022 were tennis players, with Naomi Osaka making $50.1 million and Serena Williams making $41.3 million (Knight, 2023b). This was only about half of the earnings of male tennis counterpart Roger Federer, who made $95.1 million in the same year (Knight, 2023a). As a result of the disparity in coverage of female sports, the gender pay gap will most likely endure until media coverage gains parity (Morgan, 2021; Neal, 2018). Likewise, male athletes will continue to be better known to the public, and children are likely to see male athletes as role models more often than female athletes and thus likely to associate sports as generally being more male-dominated and oriented, resulting in children emulating their idols based upon the sport with no gender reinterpretation.

The propensity for prepubescent children to emulate male athletes is compounded when the dominant sports in a society are those that are traditionally played by males, such as football (soccer), cricket, Australian rules football (AFL), American football (Gridiron), and ice hockey (NHL). Historically, these and other mainstream sports have traditionally been viewed as male domains, so male role models are copied, and coaches tend to use male sports data for coaching and training. Even though there are several female sports, many elite female athletes, and greater female participation in sports, female athletes are not promoted and publicized to the same extent as male athletes. For example, more new cricket players in Australia were female (Cricket Australia, 2022). Likewise, girls' and women's participation in the AFL has increased steadily, with females representing 31% of players in 2022 (AFL, 2022). Similarly, for ice hockey, female players account for 45.9% in Canada and 33.7% in the USA (Murphy, 2020). In other countries where ice hockey is played, female participation is much lower; for example, Russia has (13.4% participation), Finland (13.1%), and Sweden (29.9%) (Murphy, 2020). These figures for female participation in ice hockey are apparently suppressed due to a focus on recruiting male athletic talent to make money and a lack of attention to motivating young female athletes. Furthermore, it is probable that the high level of body contact between opposing players that has become accepted in ice hockey (for no good reason) might make it more intrinsically attractive to male players than to female players.

Stakeholders, including sporting bodies and governments, use composite data points to identify talent and stream athletes to maximize their potential to achieve success and for nations to excel in sports at the international level (Bergeron et al., 2015; Johnston et al., 2018). Data and algorithms are used to identify talent at a very young age and improve performance measurements to select potential elite athletes. This data is then used to develop models that identify metrics around athletes' physical traits and characteristics that make them ideal sports performers. Johnston et al. (2018) concluded that very little is known about predictors of talent in elite sports, and, as discussed above, given the male domination experienced in many sports, even less is known about predicting talent in female athletes. Other studies lend support to these findings (e.g., Lidor et al., 2009; Swann et al., 2014). In addition, a systematic review of the sports training and identification literature found that there was no clear set of variables that consistently predict future athletic performance and success (Johnson et al., 2018).

Data use for talent identification and athlete development is particularly problematic for female athletes because the data collected in sports is typically linked to male experiences and not female experiences, yet this data is used for the identification of talent and training of both male and female athletes (Curran et al., 2019; Johnston et al., 2018; McMahon & Barker-Ruchti, 2017; Ramos et al., 2019). Furthermore, collected data standards or performance indices are often drawn from data on adult athletes and are not directly applicable to young adolescent or prepubescent athletes (Ramos et al., 2019; Rubajczyk et al., 2017). Consequently, using adult and predominantly male data metrics for prepubescent athletes and younger female athletes to inform their sporting development design is most likely flawed.

In a systematic review of how sports measurements and data have been used in athlete development and talent identification, Baker et al. (2020) noted that the concept of talent identification and understanding about what to measure is still relatively unknown and not well understood. Nevertheless, data on the individual athlete and measuring athletic performance remain central to how coaches, trainers, and athletes think about performance development. Furthermore, among athletes, sports bodies, government policymakers, and coaches, it is widely believed that data can be used to identify and select individuals at an early age, which is assumed to lead to enhanced athlete development and positive sporting outcomes (Baker & Wattie, 2018; Baker et al., 2020), such as winning, increased attendance at games, and more purchases of merchandise. Baker and Wattie (2018) also identified that ineffective or inaccurate decisions connected to athlete selection and performance—which, for reasons discussed above, are much more likely to affect female athletes—could affect the athletes concerned and lead to higher dropout, decreased motivation, and reduced funding. As discussed above, the way that athletes are coached and how their bodies and performances are measured has a significant impact on their performance, sporting engagement, and overall well-being (Bangsbo, 2015; Bergeron et al., 2015; D'Isanto et al., 2019; Fonti et al., 2023; Johnston et al., 2018; Lee et al., 2021; Lidor et al., 2009; McMahon & Barker-Ruchti, 2017; McMahon & Penney, 2013; Midgley et al., 2021; Rein & Memmert, 2016).

3 Developmental Considerations and Conclusion

Within a developmental framework, it is important to have operational definitions of sport and performance, as there will always be opposition to both paradigms. In a developmental sense, all sport has a positive effect and is associated with play as an entertaining archetype; however, performance is primarily aimed at winning or mastering a game/sport (Rein & Memmert, 2016; J-F et al., 2017). Sports should be encouraged for all children, male or female. Data collection and analytics should not limit the chances for success and constructive play but should work toward increasing pleasure and minimizing injuries. However, when a child quickly demonstrates ability or excels in a sport, measures should be implemented to support and foster their development in that sport. It is at this level, developmentally, that participation in sports leads to indisputable and long-term developmental gains because sports anchor the development of milestones physically, socially, and emotionally (Eime et al., 2013; Holt et al., 2017). Athletic performance should match children's development. Thus, coaches' and athletes' strategies, decisions, and training programs should be aligned with children's developmental needs (Rein & Memmert, 2016). At the same time, sports analytics can lead to improvements in young people's physical activity, with studies showing the benefit of increasing physical activity in the younger years (e.g., McGregor et al., 2015). The microscopic to macroscopic principles for playing a sport and performing must be developmentally appropriate for the child in terms of the skills and tactics necessary for success (Rein & Memmert, 2016), thus maximizing the space for development, skill, and time for participation in constructive activity (Rein & Memmert, 2016).

Making an accurate and valid prediction about an individual's ability to perform well at a chosen sport or an athlete's long-term success in a professional sport is difficult to predict. In a study of athletes in the Canadian National Hockey League, Gee et al. (2010) examined the use of personality inventories as an aid for coaches to inform their selection decisions. Their study concluded that personality profiles could help coaches identify higher probabilities of athlete success in the future and assist players with mental health issues. In a cross-sectional study of talent identification across multiple team sports, including football (soccer), rugby union, rugby league, Australian rules football, netball, basketball, and field hockey, Barraclough et al. (2022) found that previous studies of these sports using athlete biometric data—such as GPS, BIA, flexibility, and reaction times—could lead to misrepresentations of athletes‘ potential because different physical characteristics evolve and develop at different rates for each individual. This finding is well supported by others who go further and state that biometric data has a confounding effect due to individual differences pertaining to growth, maturation, and development (Eskandarifard et al., 2022; Haugen & Seiler, 2015; Hill et al., 2021; Utczás et al., 2020). Thus, the use of inappropriate data selection that fails to recognize the developmental requirements of females and males and not using age-appropriate measures of sporting performance can deter athletic development.

Sports analytics for young people needs to be carefully considered. The authors advocate that such a sports analytic approach should use a developmental lens. The use of data analytics for positive development or to enhance one's trajectory of improvement is by no means unethical or improper, but when it is used with the preconceived notion of developing a skill or tactic at the expense of development needs (or being detrimental to them), it must be challenged and critically reviewed. Importantly, the transition from childhood to adulthood is critical regarding patterns of behavior because lifestyle choices established during adolescence have an enduring impact on health in adulthood (Lombard et al., 2017, McGregor et al., 2015).

Female participation in sports is important because female athletes provide role models and assist other girls and young women to navigate gender and sexuality norms, which are integral for girls and young women to commence and continue to participate in their sport. Having more females in sports will also eventually allow coaches and trainers access to gender-specific data that can be used to support and tailor appropriate training programs instead of relying on data derived from male athletes, which otherwise inappropriately masculinizes sport and recapitulates historically constructed male stereotypes pertaining to participation in sports. Female representation in sports and gender-specific data collection is likely to enhance girls' and women's participation and connectedness with the sport. Thus, when young people—female or male—see others like themselves participating in sports, it alters their concept of gender image, builds social connectedness, and, ultimately, encourages greater participation and continuation in the sport, all of which positively impacts on young people's physical, social, and psychological well-being.

References

Al Raisi, O. (2023). Premier league football clubs marketing costs data. *The Sports Journal*, May16, 2023. https://sportsjournal.io/premier-league-football-clubs-marketing-costs-data/

AustralianFootballLeague. (2022). *AustralianFootballLeagueAnnualReport2022*. https://resources.afl.com.au/afl/document/2023/03/03/a2ac5fc4-1ce6-4a12-85e6-9d398a960142/2022-AFL_AR_Digital-Doc_150dpi_Low.pdf

Australian Government. (2020). *Sports industry economic analysis exploring the size and growth potential of the sport industry in Australia*. https://www.health.gov.au/sites/default/files/documents/2020/05/sports-industry-economic-analysis.pdf

Australian Institute of Sport. (2023). *2032+ Australia's high performance sport strategy*. https://www.winwell2032.au/__data/assets/pdf_file/0007/1083742/Australias-2032-High-Performance-Sport-Strategy.pdf

Baker, J., & Wattie, N. (2018). Innate talent in sport: Separating myth from reality. *Current Issues in Sport Science, 3*(6). https://doi.org/10.15203/CISS_2018.006

Baker, J., Wilson, S., Johnston, K., Dehghansai, N., Koenigsberg, A., De Vegt, S., & Wattie, N. (2020). Talent research in sport 1990–2018: A scoping review. *Frontiers in Psychology, 11*, 607710. https://doi.org/10.3389/fpsyg.2020.607710

Bangsbo, J. (2015). Performance in sports with specific emphasis on the effect of intensified training. *Scandinavian Journal of Medicine & Science in Sports, 25*(Suppl. 4), 88–99. https://doi.org/10.1111/sms.12605

Barraclough, S., Till, K., Kerr, A., & Emmonds, S. (2022). Methodological approaches to talent identification in team sports: A narrative review. *Sports, 10*(6), 81. https://doi.org/10.3390/sports10060081

Bergeron, M., Mountjoy, M., Armstrong, N., Chia, M., Côté, J., Emery, C., & Engebretsen, L. (2015). International Olympic Committee consensus statement on youth athletic development. *British Journal of Sports Medicine, 49*(13), 843–851. https://doi.org/10.1136/bjsports-2015-094962

Buchheit, M., Allen, A., Poon, T. K., Modonutti, M., Gregson, W., & Di Salvo, V. (2014). Integrating different tracking systems in football: Multiple camera semi-automatic system, local position measurement and GPS technologies. *Journal of Sports Sciences, 32*(20), 1844–1857. https://doi.org/10.1080/02640414.2014.942687

Canadian-Sport-Institute-Network. (2023). *Own the podium*. https://www.ownthepodium.org/en-CA/Partners/Canadian-Sport-Institute-Network

Cricket Australia. (2022). *2021–2022 Annual Report*. https://read.e-brochures.com.au/cricketaustralia/annual report-2021-2022/pdf.pdf

Cumming, P. S., Lloyd, S. R., Oliver, L. J., Eisenmann, J. C., & Malina, M. R. (2017). Bio-banding in sport: Applications to competition, talent identification, and strength and conditioning of youth athletes. *National Strength and Conditioning Association, 39*(2), 34–47. https://sportforlife.ca/wp-content/uploads/2017/09/biobanding.pdf

Curran, O., MacNamara, A., & Passmore, D. (2019). What about the girls? Exploring the gender data gap in talent development. *Frontiers in Sports and Active Living, 1*, 3. https://doi.org/10.3389/fspor.2019.00003

D'Isanto, T., D'Elia, F., Raiola, G., & Altavilla, G. (2019). Assessment of sport performance: Theoretical aspects and practical indications. *Sport Mont, 17*(1), 79–82. https://doi.org/10.26773/smj.190214

Devlin, T. (2021). A world tour of athletics: The most popular sports by country. *+Babbel Magazine*. https://www.babbel.com/en/magazine/most-popular-sports-by-country

Eime, R., Young, J., Harvey, J., Charity, M., & Payne, W. (2013). A systematic review of the psychological and social benefits of participation in sport for children and adolescents: Informing development of a conceptual model of health through sport. *International Journal of Behavioral Nutrition and Physical Activity, 10*(98). https://doi.org/10.1186/1479-5868-10-98

Eskandarifard, E., Nobari, H., Sogut, M., Clemente, F., & Figueiredo, A. (2022). Exploring interactions between maturity status and playing time with fluctuations in physical fitness and hormonal markers in youth soccer players. *Scientific Reports, 12*(1), 4463. https://doi.org/10.1038/s41598-022-08567-5

Fonti, F., Ross, J., & Aversa, P. (2023). Using sports data to advance management research: A review and a guide for future studies. *Journal of Management, 49*(1), 325–362. https://doi.org/10.1177/01492063221117525

Gee, C., Marshall, J., & King, J. (2010). Should coaches use personality assessments in the talent identification process? A 15 year predictive study on professional hockey players. *International Journal of Coaching Science, 4*(1), 25–34.

Gitnux. (2023). *Jobs in sports 2023: Statistics and trends*. https://blog.gitnux.com/job-opportunities-in-sports-statistics/#:~:text=In%20the%20United%20States%2C%20there,opportunity%20that%20exists%20within%20it

Hamlin, M., Wilkes, D., Elliot, C., Lizamore, C., & Kathiravel, Y. (2019). Monitoring training loads and perceived stress in young elite university athletes. *Frontiers in Physiology, 10*, 34. https://doi.org/10.3389/fphys.2019.00034

Hammes, F., Hagg, A., Asteroth, A., & Link, D. (2022). Artificial intelligence in elite sports—A narrative review of success stories and challenges. *Frontiers in Sports and Active Living, 4*. https://www.ncbi.nlm.nih.gov/pmc/articles/PMC9309390/

Harell, A., & Bajic, V. I. (2020). *The data gap in sports analytics and how to close it*. Association for the Advancement of Artificial Intelligence School. https://ai-teamsports.weebly.com/uploads/1/2/7/0/127046800/paper9.pdf

Harley, J., Lovell, R., Barnes, C., Portas, M., & Weston, M. (2011). The interchangeability of global positioning system and semiautomated video-based performance data during elite soccer match play. *Journal of Strength Conditioning Research, 25*(8), 2334–2336. https://doi.org/10.1519/JSC.0b013e3181f0a88f

Haugen, T., & Seiler, S. (2015). Physical and physiological testing of soccer players: Why, what and how should we measure? *Sportscience, 19*, 10–27.

Hill, M., Scott, S., McGee, D., & Cumming, S. (2021). Are relative age and biological ages associated with coaches' evaluations of match performance in male academy soccer players? *International Journal of Sports Science & Coaching, 16*(2), 227–235. https://doi.org/10.1177/1747954120966886

Holt, L. N., Neely, C. K., Slater, G. L., Camiré, M., Côté, J., Fraser-Thomas, J., MacDonald, D., Strachan, L., & Tamminen, A. K. (2017). A grounded theory of positive youth development through sport based on results from a qualitative meta-study. *International Review of Sport and Exercise Psychology, 10*(1), 1–49. https://doi.org/10.1080/1750984X.2016.1180704

Impellizzeri, F., Marcora, S., & Coutts, A. (2019). Internal and external training load: 15 years on. *International Journal of Sports Physiology and Performance, 14*(2), 270–273. https://doi.org/10.1123/ijspp.2018-0935

Industry Europe. (2023). *European and American economic models of sport organizations*. European and American Economic Models of Sport Organizations—Industry Europe. https://industryeurope.com/sectors/politics-economics/european-and-american-economic-models-of-sport-organizations/

J-F, Pullen, D., & Carroll, A. (2013). Conceptualizing adolescent wellbeing within aneducation context. In J-F, D. Pullen, & A. Carroll (Eds.), *Adolescent wellbeing: Trends, issues and prospects*. Australian Clearinghouse for Youth Studies.

J-F, Pullen, D., Swabey, K., Carroll, A., Heath, A., Lombard, S., Yael, I.-C., O'Grady, P., & Garate, M. P. (Eds.). (2017). *Lifespan development in an educational context: (edited book)*. Wiley.

Johnston, K., Wattie, N., Schorer, J., & Baker, J. (2018). Talent identification in sport: A systematic review. *Sports Medicine, 48*, 97–109. https://doi.org/10.1007/s40279-017-0803-2

Karp, A. (2023). *SBJ marketing: Ad spending up big for NBA game telecasts*. https://www.sportsbusinessjournal.com/SB-Blogs/Newsletter-Marketing/2022/04/13.aspx

Kilic, K., & Ince, M. L. (2021). Youth athletes' developmental outcomes by age, gender, and type of sport. *Journal of Human Sport and Exercise, 16*(1), 212–225. https://doi.org/10.14198/jhse.2021.161.19

Knight, B. (2023a). The world's 10 highest-paid athletes 2023. *Forbes*. https://www.forbes.com/sites/brettknight/2023/05/02/the-worlds-10-highest-paid-athletes-2023/?sh=69e3dceeb156

Knight, B. (2023b). The world's highest-paid female athletes 2022. *Forbes*. https://www.forbes.com/sites/brettknight/2023/05/02/the-worlds-10-highest-paid-athletes-2023/?sh=69e3dceeb156. https://www.forbes.com/sites/brettknight/2022/12/22/the-worlds-highest-paid-female-athletes-2022/?sh=d74b21f73b49

Lee, S., Kwon, S., & Ahn, J. (2021). The effect of modeling on self-efficacy and flow state of adolescent athletes through role models. *Frontiers in Psychology, 12*, 661557. https://doi.org/10.3389/fpsyg.2021.661557

Leng, H., & Phua, Y. (2022). Athletes as role models during the COVID-19 pandemic. *Managing Sport and Leisure, 27*(1–2), 163–167. https://doi.org/10.1080/23750472.2020.1762330

Lerner, R. M., Brown, J. D., & Kier, C. (2005). *Adolescence: Development, diversity, context, and application*. Pearson Education Canada.

Lichtenthaler, U. (2021). Mixing data analytics with intuition: Liverpool football club scores with integrated intelligence. *Journal of Business Strategy, 43*(1), 10–16. https://doi.org/10.1108/JBS-06-2020-0144

Lidor, R., Côté, L., & Hackfort, D. (2009). To test or not to test? The use of physical skill tests in talent detection and in early phases of sport development. *International Journal of Sport and Exercise Psychology, 7*, 131–146.

Lombard, S. F. J., Pullen, D., & Swabey, K. (2017). Conceptualizing trauma for children of drug addicted mothers: A developmental mapping. *Journal of Trauma & Treatment, 6*(1). https://doi.org/10.4172/2167-1222.1000352

McGregor, E., Swabey, K., & Pullen, D. (2015). How often do you move? Improving student learning in the elementary classroom through purposeful movement. *Open Journal of Social Sciences, 3*(6), 6–10. https://doi.org/10.4236/jss.2015.36002

McMahon, J., & Barker-Ruchti, N. (2017). Assimilating to a boy's body shape for the sake of performance: Three female athletes' body experiences in a sporting culture. *Sport, Education and Society, 22*(2), 157–174. https://doi.org/10.1080/13573322.2015.1013463

McMahon, J., & Penney, D. (2013). (Self-)surveillance and (self-)regulation: Living by fat numbers within and beyond a sporting culture. *Qualitative Research in Sport Exercise and Health, 5*(2), 157–178. https://doi.org/10.1080/2159676X.2012.712998

Midgley, C., DeBues-Stafford, G., Lockwood, P., & Thai, S. (2021). She needs to see it to be it: The importance of same-gender athletic role models. *Sex Roles, 85*, 142–160. https://doi.org/10.1007/s11199-020-01209-y

Morgan, S. (2021). Working twice as hard for less than half as much: A socio-legal critique of the gendered justifications perpetuating unequal pay in sports. *The Columbia Journal of Law & the Arts*, April 5. https://ssrn.com/abstract=3819943

Moura, F., Martins, L., Anido, R., de Barros, R., & Cunha, S. (2012). Quantitative analysis of Brazilian football players' organisation on the pitch. *Sports Biomechanics, 11*(1), 85–96. https://doi.org/10.1080/14763141.2011.637123

Murphy, M. (2020). The prevalence and growth of women's hockey across the globe. *The Ice Garden*, (July, 17). https://www.theicegarden.com/the-prevalence-and-growth-of-womens-hockey-across-the-globe-iihf-player-registration-data/#:~:text=Today%2C%20there%20are%20209%2C966%20registered,in%20the%20U.S.%20and%20Canada

Neal, A. (2018, July 13). Yes, the WNBA wage gap is a real thing. *Grandstand Central*. https://grandstandcentral.com/2018/sports/basketball/yes-the-wnba-wagegap-is-real/

Opstoel, K., Chapelle, L., Prins, F., De Meester, A., Haerens, L., van Tartwijk, J., & De Martelaer, K. (2020). Personal and social development in physical education and sports: A review study. *European Physical Education Review, 26*(4), 797–813. https://doi.org/10.1177/1356336X19882054

Patel, D., Shah, D., & Shah, M. (2020). The intertwine of brain and body: A quantitative analysis on how big data influences the system of sports. *Annals of Data Science, 7*, 1–16. https://doi.org/10.1007/s40745-019-00239-y

Ramos, S., Volossovitch, A., Ferreira, A. P., Fragoso, I., & Massuça, L. (2019). Differences in maturity, morphological and physical attributes between players selected to the primary and secondary teams of a Portuguese basketball elite academy. *Journal of Sports Sciences, 37*(15), 1681–1689. https://doi.org/10.1080/02640414.2019.1585410

Reid, H. (2017). Athletes as heroes and role models: An ancient model. *Sport, Ethics and Philosophy, 11*(1), 40–51. https://doi.org/10.1080/17511321.2016.1261931

Rein, R., & Memmert, D. (2016). Big data and tactical analysis in elite soccer: Future challenges and opportunities for sports science. *Springer Plus, 5*, 1410. https://doi.org/10.1186/s40064-016-3108-2

Rubajczyk, K., Świerzko, K., & Rokita, A. (2017). Doubly disadvantaged? The relative age effect in Poland's basketball players. *Journal of Sports Science and Medicine, 16*(2), 280–285.

Singh, B., Olds, T., Curtis, R., Dumuid, D., Virgara, R., Watson, A., & Maher, C. (2023). Effectiveness of physical activity interventions for improving depression, anxiety and distress: An overview of systematic reviews. *British Journal of Sports Medicine, 57*(18), 1203–1209. https://doi.org/10.1136/bjsports-2022-106195

Stålstrøm, J., Iskhakova, M., & Pedersen, Z. (2023). Role models and athlete expression at the youth Olympic games as impactful sport communication practices. *International Journal of Sport Communication, 1-15*, 435. https://doi.org/10.1123/ijsc.2022-0212

Swann, C., Moran, A., & Piggott, D. (2014). Defining elite athletes: Issues in the study of expert performance in sport psychology. *Psychology of Sport and Exercise, 16*, 3–14. https://doi.org/10.1016/j.psychsport.2014.07.004

Teigen, K., Normann, H., Bjorkheim, J., & Helland, S. (2000). Who would you most like to like? Adolescents' ideals at the beginning and the end of the century. *Scandinavian Journal of Educational Research, 44*(1), 5–26. https://doi.org/10.1080/713696661

Utczás, K., Tróznai, Z., Pálinkás, G., Kalabiska, I., & Petridis, L. (2020). How length sizes affect body composition estimation in adolescent athletes using bioelectrical impedance. *Journal of Sports Science & Medicine, 19*(3), 577.

Wenner, L. (2021). Media, sports, and society. In E. Pike (Ed.), *Research handbook on sports and society* (Vol. 27, pp. 111–126). Edward Elgar Publishing. https://doi.org/10.5860/choice.27-4339

Willson, E., & Kerr, G. (2022). Body shaming as a form of emotional abuse in sport. *International Journal of Sport and Exercise Psychology, 20*(5), 1452–1470. https://doi.org/10.1080/1612197X.2021.1979079

World Health Organization. (2021). *Physical activity fact sheet* (No. WHO/HEP/HPR/RUN/2021.2). World Health Organization. https://apps.who.int/iris/bitstream/handle/10665/346252/WHO HEP-HPR-RUN-2021.2-eng.pdf?sequence=1

Decision Intelligence in Sports Marketing

P. Mary Jeyanthi, Violeta Cvetkoska, and Bojan Kitanovikj

1 Introduction

Decision intelligence in sports marketing refers to the use of advanced analytics and decision-making tools to inform strategic decisions related to sports marketing initiatives. It involves the collection, analysis, and interpretation of data to gain insights into consumer behavior, market trends, and the effectiveness of marketing campaigns (Bukstein, 2016). By leveraging decision intelligence, sports organizations can optimize their marketing strategies, improve fan engagement, and drive revenue growth. Key applications of decision intelligence in sports marketing include customer segmentation, personalized marketing, pricing optimization, and social media analytics (Apostolou & Tjortjis, 2019). Considering big data (BD) has become a vital tool in contemporary research and sports management practices, managers, business leaders, and marketers have all the information they need to make informed decisions efficiently (Watanabe et al., 2021).

Yet, to make sense of the data, they often rely on artificial intelligence (AI) and related technology, thus contributing to the rise of the new academic discipline that blends engineering, social sciences, and management and decision theories called decision intelligence (Duan et al., 2019). AI is not a stranger to the sports industry. From match outcome predictions and strategic and tactical decision-making to fantasy sports injury predictions, AI has crept into sports at various degrees of application and acceptance (Beal et al., 2019). Recent projections regarding the global AI in the sports market point out that it could reach a value of 19.2 billion USD with an

P. M. Jeyanthi (✉)
Jaipuria Institute of Management, Jaipur, Rajasthan, India

V. Cvetkoska · B. Kitanovikj
Faculty of Economics-Skopje, Ss. Cyril and Methodius University in Skopje, Skopje, North Macedonia
e-mail: vcvetkoska@eccf.ukim.edu.mk; bojan.kitanovikj@eccf.ukim.edu.mk

A. Mansurali et al. (eds.), *Sports Analytics*,
https://doi.org/10.1007/978-3-031-63573-1_3

annual growth rate of 30.3% until 2030 and highlight machine learning (ML) as one of the most lucrative segments on the market (Beesetty et al., 2022). This opens the door for a wider application of AI technologies in nurturing decision-intelligence practices in the sports industry. One of them is sports marketing, as one of the fastest-growing aspects of sports (Fullerton & Merz, 2008). As a result, this chapter aims to explore the role of decision intelligence in sports marketing with an emphasis on the use of predictive analytics.

Even though professional and widely popular leagues and clubs across different sports like baseball, basketball, football, and the like use big data and contemporary analytics in marketing, there is still a wide gap in the sports marketing community regarding theoretical knowledge, applications, and best practices. An effort is made through this book chapter to bridge that gap and offer a synthesized approach to how decisions are made in sports marketing with the use of business analytics.

The chapter presents a detailed overview of the existing literature and the related work explaining the current stage of decision intelligence as well as sports marketing, through a concise evolutionary framework and a summary of common practices and findings. Afterward, the authors introduce the readers to decision intelligence and sports marketing and the possibility of them going hand in hand to boost sales and growth for sports teams and related organizations in the industry. The chapter goes on to emphasize the application of predictive analytics in sports marketing to determine the likelihood of specific future events happening. Despite descriptive analytics, the power of sophisticated data analysis lies in predictive analytics that can model possible fan behavior to renew sports season tickets, likely fluctuations of prices, factors that drive attendance at sports games, and more (Hensley, 2022).

Being aware of this information has the potential to significantly improve the productivity of sales departments of sports organizations, enhance the effectiveness of marketing efforts, lead to increased efficiency of the use of resources, and ultimately drive profits (Mumcu & Fried, 2017). Predictive analytics in sports marketing is possible with numerous techniques and AI-driven algorithms, which come up with decision alternatives, fostering a decision intelligence climate for members of the management and marketing teams. Based on the presented data, the chapter will outline the next steps and future paths of research and practical application of decision intelligence in sports marketing and the process of making informed decisions aided by data. This will be followed by a set of conclusions for researchers, managers, and marketers, thus contributing to a deeper understanding of how decision intelligence is applied in sports marketing, most notably for predictive analytics of likely outcomes.

Furthermore, the purpose of this book chapter is to highlight the importance of decision intelligence and predictive analytics as powerful tools for professionals in sports marketing, enabling them to make data-driven decisions for boosting fan engagement through social media and ticket sales analyses, optimizing revenue streams, and improving marketing campaigns and the overall competitiveness and organizational growth. Through a review of the existing literature and the opportunities presented by predictive analytics, both researchers and practitioners in sports marketing can get new insight into where decision intelligence is going in the future and how they can benefit from applying decision intelligence techniques to boost their marketing performance.

2 Literature Review

When it comes to sports, the marketing approaches vary when compared to other sectors. Moreover, sports marketing incorporates two main components. They are related to offering sports services and products to the customers and shifting other industrial and consumption services and products through the promotional aspect of sports (Dees et al., 2022). Aside from this, the differences between sports marketing approaches and other sectors are derived from numerous reasons that are characteristic of sports marketing (Dees et al., 2022).

It can be clearly understood that sports marketing characteristics are pronounced by the analysis of its overall features, its market features (Dees et al., 2022), communication on a marketing level, marketing mix, and consumption (da Silva & Las Casas, 2020). It is due to these various features that sports marketing is strategically placed and dealt with.

Furthermore, the entire organization is involved in the strategic concept that has an adaptive and flexible framework. Strategic marketing techniques are used by marketing professionals, academics, and executives so that firms may gain a lasting advantage over the competition (Snyder et al., 2016). This leads to the initiation of a method for developing a structural strategy and planning (Fan et al., 2015). The idea of marketing intelligence has emerged as a result of this strategic marketing viewpoint.

Marketing intelligence, according to Dam et al. (2019), is the capacity to understand, analyze, and evaluate information about an organization's clients, market, competitors, and other businesses, which is obtained from both the internal and external surroundings. This information is used to enhance strategic and tactical decision-making. Additionally, they underline that marketing analysis, research, and the study of financial and business data are all components of competitive intelligence (Dam et al., 2019). This concept incorporates the strategic elements of marketing intelligence covered by Venter and Jansen van Rensburg (2014). Marketing intelligence, in this perspective, is essential to decision-making and aids in information sharing inside organizations (Fan et al., 2015).

According to related research and literature, marketing intelligence needs to be simple for recipients to understand it and it should take the shape of an incentive (Dam et al., 2019). Additionally, the foundation of marketing intelligence is based on trust. The sender should obtain the recipient's trust, and the utilized information should be connected with the recipients in a trustworthy way (Watanabe et al., 2021). Even though Venter and Jansen van Rensburg (2014) explore the concept of marketing intelligence within the context of marketing research, this aspect is evaluated by concentrating on strategy beyond inquiry which is included within the marketing research framework (Dewi & Darma, 2019). While businesses use marketing research when they need specialized data or when it represents a response to a specific set of their needs, intelligence is portrayed as a perpetual process with a holistic approach to the business surroundings, including customers, competitors, and the market (Dam et al., 2019).

For some other authors, marketing intelligence is the most recent knowledge gleaned through observations of people's attitudes, actions, and demands about the environment, society, and everyday economic activity (Bukstein, 2016). In terms of the ultimate objective of marketing intelligence, it lies in facilitating a decision-making process that results in an action (Venter & Jansen van Rensburg, 2014). The effectiveness of intelligence processes builds organizational knowledge and promotes cumulative learning (Dam et al., 2019). According to Yeoh and Popovič (2016), marketing intelligence attempts to extract knowledge out of data and intelligence out of knowledge.

Having all of this in mind regarding marketing intelligence, the notions of sports marketing and sports marketing intelligence must be expressed from the viewpoint that marketing intelligence may develop into a commercial competence to offer a competitive advantage. Nowadays, sports marketing intelligence is used to methodically gather and analyze data to leverage such knowledge to advance the objectives of strategic management or marketing. In addition, it emphasizes technical, social political, and economic advancements as well as the competitive surroundings (Dewi & Darma, 2019).

More precisely, sports marketing intelligence encompasses the ability of organizations to analyze, understand, identify, evaluate, and reinforce the entire internal and external surroundings and generate dynamic outcomes in the brand formation process to enhance strategic and tactical decision-making (Ranjan & Foropon, 2021). Sports marketing intelligence is a key component of a company's sustainability and a key indicator of its competitive advantage (Katsikea et al., 2019). It has three components: gathering data, implementing possibilities, and formulating a strategy that will work (Dewi & Darma, 2019). These aspects of sports marketing intelligence are intended to assist in gathering accurate data across a broad spectrum, understanding the market structure, consumer actions, and expectations, and transforming them into a distinctive or different ability of organizations, thereby generating the future perspective. The main goals of sports marketing intelligence are to implement best practices and enhance corporate performance (Goebert & Greenhalgh, 2020).

This strategy, which includes businesses' future viewpoints, may be implemented by creating competition possibilities, reducing or removing any shocks that may arise, giving organizations an edge in terms of response time, and seizing opportunities (Goebert & Greenhalgh, 2020). The capacity to build brand identity, which is considered to be a sign of sports marketing intelligence, expresses this integrated framework.

Another similar concept in sports marketing is decision intelligence, which represents a broader paradigm of making decisions in various spheres and core functions of an organization. Duan et al. (2019) define decision intelligence as the implementation of data-driven business analytics and technologies based on AI and machine learning to enhance and improve decision-making. The end goal is for people to be able to make quicker and more relevant decisions instead of fully replacing the human touch and intervention in the decision-making process.

Advanced, emerging technologies in sports marketing can also help with collecting, cleaning, analyzing, and presenting data, as well as creating interactive dashboards. Sports marketing nowadays deals with Big Data, which can comprehensively give an overview of the history of prices, patterns, and behaviors of fans, social media presence and engagement, and so on (Mamo et al., 2022). For the latter, sports marketers use advanced social media analytics to track their performance on these platforms, including brand reputation and sentiment analysis (Vale & Fernandes, 2018). To process this data, sports marketers use tools based on certain programming languages like Python, as well as AI and ML. AI's capabilities continue to be a relatively unexplored frontier. Its combination with the Internet of Things (IoT) can help sports organizations automate customer relationship management (CRM) and ethically keep track of sports fans' data (Zhang & Zhu, 2022). Additionally, instead of just showcasing the present situation, business intelligence finds a way to help organizations predict revenue, fan engagement, and game attendance, with predictive analytics software (Hubáček et al., 2019).

3 Introduction to Decision Intelligence in Sports Marketing

As stated, decision intelligence is a field stemming from decision-making theories and applications that combine AI, data analytics, and human expertise for choosing the best most suitable alternatives and ultimately making informed and effective business decisions. For the purpose of this book chapter, it is observed in the context of sports marketing where it can impact the optimization of marketing strategies, and the efforts to boost fan engagement, and revenue generation. In other words, by applying different decision intelligence and data analytics techniques, sports marketers can get new insights into the behaviors and preferences of the fanbase and observe market trends that can serve as inputs for data-driven decision-making.

To begin with, decision intelligence can be employed for fan segmentation. Furthermore, sports marketers can segment their fan bases into separate groups, which have distinctive characteristics and behaviors based on factors such as buying patterns, engagement on social media, demographics, history of attending sports games, and similar. Illustratively, a sports marketing team may identify a special segment of fans who are more likely to attend games with their families. With this information, the team can create targeted marketing initiatives like child-friendly events, discounted ticket packages for families, and others in an effort to retain these fans. That way, marketers can ensure higher engagement and conversion rates on their marketing campaigns and messaging, too.

N3XT Sports (n.d) is a case study of a company that is assisting sports organizations in dividing their over-the-top (OTT) subscriber database into different segments (Table 1). In this case, the fan segmentation process involves the categorization

Table 1 Sample fan segregation

Type of fan	Criteria	Frequency	Results
Real fan	Registered for the OTT	Watches weekly for the past three months	Exclude people who unsubscribed
Engaged fan	Registered for the OTT, yet watched more than five times	Once in a month	Exclude people who unsubscribed
Light fan	Registered for the OOT, yet watched once or no more than five times	Once in a month	Exclude people who unsubscribed
Curious fan	Logged in or registered for the OTT but never watched a video	More than once	Exclude people who unsubscribed
Interested fan	Their email address in the database, but are not a member of the OTT	Once	Exclude people who unsubscribed

Source: Adapted from N3XT Sports (n.d.)

of subscribers to the service based on the frequency of watching the content. Afterward, the sports marketing team can use these categories as stages of the fan journey to develop new approaches tailored to each group.

Pricing as one of the elements of the marketing mix is in the domain of sports marketing, considering that ticket prices can be the backbone of the revenue of many sports organizations. Considering the price elasticity of different identified fan segments, sports marketers can attempt to find the most suitable level of ticket prices to maximize revenue without negatively affecting fan attendance during the games. As a result, decision intelligence can help optimize pricing strategies through the analysis of key factors like historical attendance data, demand forecasts, patterns of ticket buying, and competitors' prices.

Another key area where decision intelligence can help sports marketers is the personalization of marketing initiatives. In an effort to nurture deeper connections with the fans, deliver a special fan experience, and increase fan loyalty, sports marketers opt for personalization. An example of that is customized marketing campaigns, which can be developed after a thorough analysis of individual fan data, ticket buying patterns, reactions, and engagement on social media, and similar. In this sense, a sports apparel brand can apply decision intelligence techniques to send personalized emails, newsletters, and notifications to fans who have previously shown interest in a particular player. By offering exclusive discounts or limited-edition merchandise related to that player, the brand can enhance fan engagement and improve sales.

Besides sports players, another important stakeholder in sports organizations is the sponsors. These organizations rely on them to generate revenue. So, decision intelligence can help in identifying the most suitable sponsors by analyzing data on sponsor preferences, target audience alignment, and return on investment (ROI) metrics. By understanding the value and impact of different sponsorship opportunities, sports marketers can make informed decisions about partnerships that deliver mutual benefits for both the brand and the sports entity. For instance, a basketball team can leverage decision intelligence to analyze potential sponsors' target demographics and brand positioning. By identifying sponsors whose target audience aligns with the team's fan base, the team can negotiate favorable sponsorship deals that maximize exposure and revenue generation.

4 Predictive Analytics in Sports Marketing

Data is more effectively defined as big data when its features provoke a person to alter their perspective (Tonidandel et al., 2018). Considering this, it is critical to investigate how big data may affect sports marketing and the researchers' approaches to scanning, recognizing, obtaining, and processing data to handle difficult problems and add to the body of knowledge.

While still having physical education at its heart, research on sports marketing also emphasizes this industry's business actions (Dees et al., 2022). In turn, these business activities have facilitated the need for an increase in digital technology use by sports teams, sports federations and non-profits, international sports rule-making organizations, and similar. Through digital technology, these members of the sports ecosystem interact and engage with sports fans and customers (Su et al., 2020). This makes sports marketing, and sport in general, a data-rich environment where both structured and unstructured data are generated from the digital communication channels of customers, fans, and sport-related organizations and the content all of these stakeholders produce (Wedel & Kannan, 2016).

With that in mind, big data in sports relates to the process of identification and analysis of diverse data types from data-rich environments with the purpose of decision-making in athletic actions, sport-associated business activities, and products or services (Watanabe et al., 2021). In other words, data in sports marketing can come in various degrees of structure, can be continuously and heterogeneously added, and is integral to making decisions with internal and external implications.

To make sense of sport-related data, business analytics represents the process of data transformation, using insights from previous business performance to propel business planning and decision-making forward (Mamo et al., 2022). Oftentimes, how well an organization performs in business analytics means how far ahead it is in front of the competition (Cvetkoska, 2022).

Business analytics rests on four pillars, including descriptive, diagnostic, predictive, and prescriptive analytics shown in Fig. 1 (Power et al., 2018). While the first two types of analytics focus on historical data based on the hindsight of the analysts, their outputs are tasked with providing answers to which activities occurred (descriptive analytics) as well as the reason behind these occurrences (diagnostic analytics) (Power et al., 2018). A lot of domains of sports marketing and practice tend to explore sets of patterns and identify correlations to get a broader view of sports organizations' activities and people without preconceived structure as part of exploratory analyses and descriptive analytics (Evans, 2020). To illustrate, Chang

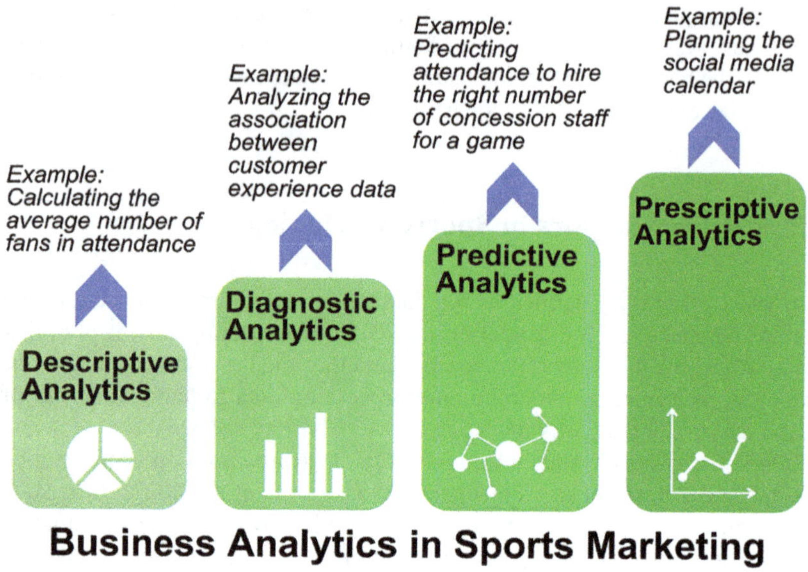

Fig. 1 Pillars of business analytics in sports marketing. (Source: Adapted from Houtmeyers, K. C., Jaspers, A., & Figueiredo, P. (2021). Managing the training process in elite sports: from descriptive to prescriptive data analytics. *International Journal of Sports Physiology and Performance*, 16(11), 1719–1723)

(2019) showcased how the feelings of fans change on Twitter during a Super Bowl 50 game based on their location and period.

The third and fourth pillars of business analytics consider providing the answer once the reasons are identified (predictive analytics) and the courses of action that should be taken to achieve the planned state (prescriptive analytics) (Power et al., 2018). While the techniques of descriptive analytics include text mining, model clustering, ratio analysis, and visualization of data (Balducci & Marinova, 2018), the functions of predictive analytics take the analysis to different heights. Predictive analytics encompasses a variety of techniques (Fig. 2), which are relevant to the sports industry, including forecasting, regression analysis, machine learning, data mining, and data visualization (Cvetkoska, 2022).

Qualitative forecasting using the Delphi method represents a qualitative method often used when the data for a particular problem may be incomplete and when researchers want to forecast the probability of events that will potentially happen in the future and gauge how they affect the researched problem (Fink-Hafner et al., 2019). The method envisions engaging a panel of experts that answers sets of surveys as each set represents a new round (Fink-Hafner et al., 2019).

Besides the method's wide application in healthcare and education, it can be applied in sports marketing, too. Abdolmaleki et al. (2018) used the Delphi method to assess how the marketing mix and psychological, individual, social, and cultural factors impact sports consumer behavior.

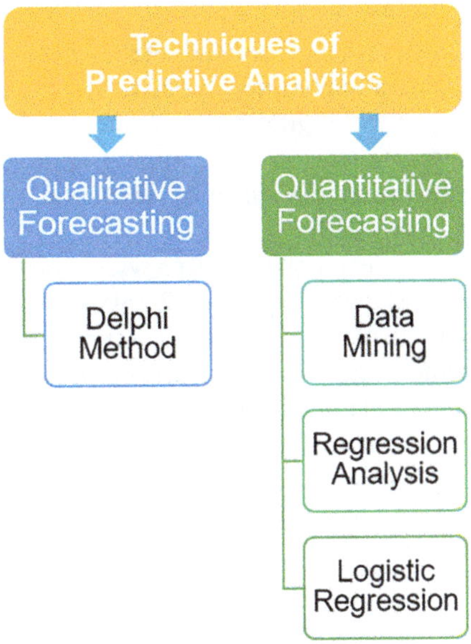

Fig. 2 Techniques of predictive analytics in sports marketing. (Source: Adapted from Apostolou, K., & Tjortjis, C. (2019, July). Sports Analytics algorithms for performance prediction. In *2019 10th International Conference on Information, Intelligence, Systems and Applications (IISA)* (pp. 1–4). IEEE)

Quantitative forecasting is used for forecasting future events based on time-series data or sequences of historical data such as prices of tickets for sports matches or sports fan likes and comments on social media. Generally, time-series data displays (1) trends and cyclical factors in which case methods like seasonal and multiplicative models, double-moving average, exponential smoothing, and Holt-Winters additive are used and (2) random behavior without any notable trends or cycles in which case researchers make forecasting with widely used statistical methods like exponential smoothing and moving average as the time series is stationary (Evans, 2020).

Jensen et al. (2017) used event history analysis as a quantitative forecasting method for analyzing how long sponsorships last during the Olympic and World Cup seasons, including a portion of sponsors who renew the sponsorship, the factors impacting the likelihood of the sponsorships to continue or not, and similar.

Regression analysis is a tool used for prediction through investigations of the relations between variables, thus forecasting a dependent variable's value with an independent variable or several of them (Guerard & Guerard, 2013).

In the context of sports marketing, performing a regression analysis, researchers have found that sports media, advertising, sponsorships, and star athletes as elements of sports marketing influence audiences' decision to follow less popular sports (Sedky et al., 2022). Additionally, regression analysis as a technique was one of the most commonly used parametric statistical techniques from 1992 to 2002 in the academic journal Sports Marketing Quarterly (Quarterman et al., 2005).

Data mining is another forecasting technique, whose utilization mainly revolves around the identification of hidden patterns, relationships between data, and the prediction of the result (Evans, 2020). In popular literature, data mining is associated with similar terms like knowledge discovery in data, machine learning algorithms, and artificial intelligence. Yet, this technique is growing in popularity in business analytics in general as well as in sports marketing considering that the majority of digital data nowadays comes in an unstructured form. To make sense of this data, several data mining approaches have been developed, including data exploration, classification, reduction, cause-and-effect models, and data association (Evans, 2020).

This technique has been widely proven applicable in sports marketing in the past. For instance, Chen et al. (2009) used data mining and the multinomial probit model, in particular, to forecast what types of tickets would single-game sports fans buy for home games based on their previous behavior of ticket buying; they found that lateness of the season, promotion, a striking opponent, and perceived value were some of the driving factors.

Logistic regression represents a powerful classification method that is used to classify the new observation into one of the previously determined classes (Cvetkoska, 2022). The dependent variable is categorical and can be transformed into a binary variable. With the logistic regression, we estimate the likelihood of each new observation and compare it with a previously set probability, so if the estimated probability is higher than the set, the observation will be in class 1 (yes action), while if it is less than the set probability, it will be in class 0 (no action) (Cvetkoska, 2022). For instance, logistic regression may be applied to determine the probability of a sports fan having an all-season pass ticket (they may have a ticket, or they may not). Prasetio and Harlili (2016) have applied logistic regression for forecasting results of football matches for an away win or a home win, based on data from training for the seasons 2010/2011 until 2015/2016 as well as the FIFA video game, and variables gathered from past research.

The majority of the predictive approaches in the past utilized in the sport-marketing-related literature apply conventional statistical techniques. Classical statistical approaches are widely used in testing existing theories and their improvements within current research paradigms (Putka et al., 2018). On the other hand, a large portion of these approaches do not practically consider nonlinearity, matching flexibility and robustness, and overfitting, and do not scale longitudinal or dynamic measurement (Yarkoni & Westfall, 2017). They frequently make the linearity assumption and are developed to organize data and at the same time suit small samples; these methods then concentrate on crunching a small number of variables to ensure a model supports and fits the vision or theory similarity explanation (Mamo et al., 2022).

Nowadays, as algorithms and big data become ubiquitous, analytical techniques are flooded with fresh opportunities for improved identification as well as model robustness. Contrary to classical analytical techniques, contemporary ones tend to be superior to them due to their capabilities to develop more advanced, efficient, and effective forecasting models, which can operate with smaller samples to a set of predictor ratios (Putka et al., 2018).

To illustrate, if someone wants to research the driving forces influencing people's attitudes, behavior, and facilitating conditions to purchase a certain sports product or service, previous research pathways in sports marketing would tilt toward a quieter approach, restricted number of predictors founded on a single theoretical framework or parts of it. A more traditional approach may consider a small portion of variables for model creation, neglecting a plethora of potential predictors like sociocultural factors, bias, individual perception, motivation, and similar, which can result in an analysis that undervalues and underuses variables, thus contributing to sports marketing theories to remain fairly simple (Putka et al., 2018).

When compared to traditional approaches, modern techniques for predictive analytics in sports marketing have the upper hand, as they are associated with enhancing accuracy, comprehensiveness, and timeliness in forecasts and classifications (Mamo et al., 2022). Another proof of that is the research on heuristics to enhance sport management's capacities to carry out management programs for ticket holders of the game season, which indicated that the random forest technique and the support vector machines as machine learning algorithms proved more accurate than classical logistic regression by 40% and 19%, respectively (Mamo et al., 2022). The bottom line is that the forecasts produced by big-data methods are more precise than classical methods like ordinary least squares regression.

Additionally, predictive analytics in sports marketing goes as far as trying to predict the amount of sold tickets and TV viewership based on big social data. The predictive model used by Egebjerg et al. (2017) performed the best when it very accurately forecasted how many attendees would be present at a football match, utilizing data from Facebook generated during a match as well as social media activity from the two weeks before the match. Considering predictive analytics is multi-faceted, Egebjerg et al. (2017) also found that since an increased social media engagement results in a higher number of spectators, sports organizations (or associations in this case) could benefit from revamping their social media marketing strategies. This can impact the type of social media posts that organizations should publish to make better returns on their posts and increase reach, traffic, and engagement, which would ultimately result in a boost in ticket sales and TV ratings.

5 Applications of Decision Intelligence in Sports Marketing

Sports organizations have used a variety of analytic models in decision-making for more than half a century in various forms and fashions. It wasn't until recently that decision intelligence in the sports business industry stopped focusing only on the player and the strategies of the sports teams on the playing field. One of the first studies on analytics in sports by Lindsey (1959) focused on an analysis of baseball players and playing strategies, including aspects related to selecting pitchers, strategically modifying lineups, determining batting averages, choosing defensive strategies, and similar. Since then, the application of decision intelligence and business analytics have deeply penetrated other layers of sports, including marketing.

Analytic teams in sports marketing have started implementing models for variable and dynamic ticket or room pricing, determining product or service availability as a result of consumer loyalty and reward programs—all of which were originally used in the airline and hospitality industries (Davenport, 2014; Stein, 2014). Sports organizations have also adopted best practices on data-driven marketing from e-commerce businesses like Amazon for tailoring digital marketing and brand loyalty strategies (website optimization, targeted email campaigns, subscriptions, targeted product/service recommendations) (Davenport, 2014).

At the start of the 2010s, pioneering sports organizations massively began integrating data-driven decision-making into strategic business functions—a majority connected to their marketing efforts. These include ticket pricing, valuation of corporate partnerships and sponsorships, customer relationship management, fan engagement, and more (Troilo et al., 2016). Nowadays, we can highlight a couple of pillar applications of decision intelligence in sports marketing (Fig. 3), including ticket pricing and sales inventory, managing customer relationships and engaging fans, digital marketing analytics and social media, market research, sponsorship valuation, retention of season ticket members, and data visualization (Bukstein, 2016).

When it comes to decision-making related to ticket inventory and prices, a vast majority of sports teams are led by maximizing spectator attendance, optimizing revenue, and creating unique value and experience for the spectators (Drayer et al., 2012). For instance, the Cleveland Indians in the USA wanted to find out the optimal mix of ticket offerings and pricing that would bring the most value to the fans and the club, resulting in launching a new product/service, which in this case included new ticket options (Gershenfeld, 2015). Furthermore, based on the prices of the tickets, the location of the seats, and the food and beverage offerings, the club's analytics team managed to increase the revenue by 6% in 2012 due to changes in the pricing strategy (lower ticket process and an increased attendance)

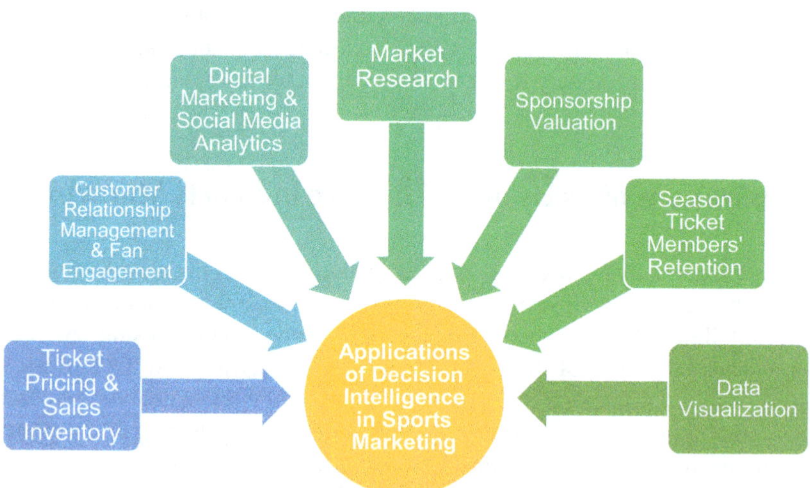

Fig. 3 Domains for applying decision intelligence in sports marketing. (Source: Adapted from Bukstein, S. (2016). Evolution and impact of business analytics in sport. In C. K. Harrison & S. Bukstein (Eds.), *Sport Business Analytics* (pp. 25–46), Auerbach Publications, Boca Raton, FL)

(Gershenfeld, 2015). Other pricing strategies like dynamic and real-time pricing based on data have been utilized by major sports leagues like the NBA, MLB, NHL, MLS, and similar (Diehl et al., 2016).

Social media as an integral part of sports marketing plays a vital role in developing trust and closeness with potential customers and fosters brand recognition. This is especially the case with young people. In a survey regarding social media use for sports purposes, almost 60% confirmed that social platforms are the place where they interact and get recommendations, whereas 43% use social media to find out about new products and services and access exclusive deals and discounts (Albán Bartra et al., 2021). Albán Bartra et al. (2021) have also demonstrated a positive impact of sports marketing strategies on creating customer value.

Forecasting sponsorship costs is aided by decision intelligence, too. Variables like on-field performance, spectators' demand for a particular sponsored property, and attendance can act as predictors of sponsorship costs (Jensen et al., 2016). Moreover, this has been found in Nike's case of sponsoring intercollegiate athletic programs. Advanced prediction models for sponsorship costs have been used by the sports apparel industry leader Nike to use its dominance in the market to negotiate the costs and make informed, data-driven decisions with sponsored properties (Jensen et al., 2016).

For the successful application of decision intelligence in sports marketing, the right set of skills is essential. Ermakov et al. (2022) champion the need for high-demand competencies in digital technologies in all spheres of sport and physical education, claiming that everyone from sports referees and coaches to athletes and sport managers, agents, and marketers should possess skills in analytics, operations, software administration, data processing, content creation, and similar. They also highlight the rising disciplines of social media marketing, artificial intelligence, and machine learning as hotspots in sports marketing (Ermakov et al., 2022).

Applying decision intelligence in sports marketing comes with its onset of challenges (Fig. 4). For one, to achieve a competitive advantage, sports management

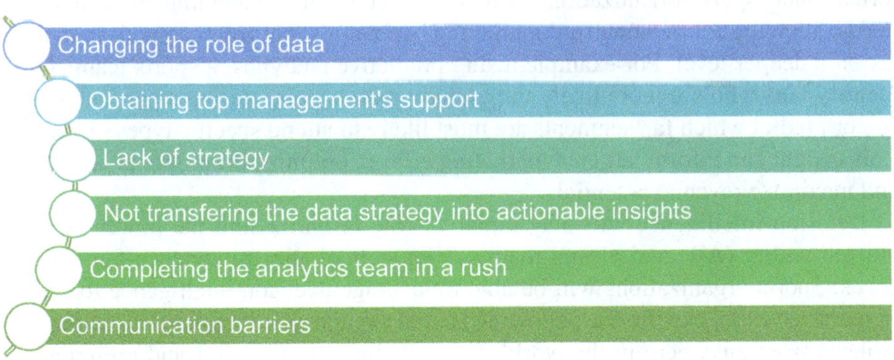

Fig. 4 Challenges in applying decision intelligence in sports marketing. (Source: Adapted from Ransbotham et al. (2016); Marr (2015); Stein and Greenland (2014); Morgulev et al., (2018); Snyder et al. (2016))

may not understand the need to change the role of data in decision-making into a more strategic one, thus holding back support (Ransbotham et al., 2016). Sports organizations may lack a strategy for implementing decision intelligence in sports marketing (Marr, 2015). On the other hand, there are cases when what is written on the paper does not translate into practice, so analytics insights are not turned into action (Stein & Greenland, 2014).

Some challenges are linked with the analytics teams, too. As sports organizations scramble to complete their analytics teams, this can result in overly complicated technological infrastructure, which is loosely and swiftly "stitched together" (Morgulev et al., 2018). Application barriers can appear due to communication gaps and a lack of effective communication within the analytics teams as well as between this team and others like the sales team of the sports organization (Snyder et al., 2016).

6 Future of Decision Intelligence in Sports Marketing

The future of decision intelligence in sports marketing holds tremendous potential for revolutionizing the way sports organizations engage with fans, optimize marketing strategies, and drive revenue generation. As technology continues to advance and data becomes increasingly abundant, decision intelligence will play a crucial role in transforming sports marketing practices. In the following paragraphs, the key areas where decision intelligence will shape the future of sports marketing are identified and elaborated.

Predictive analytics will continue to be further implemented and developed to fit the objectives of sports marketing. Decision intelligence will become even more sophisticated in leveraging predictive analytics to forecast fan behavior, market trends, and revenue opportunities. Advanced algorithms and machine learning techniques will enable sports marketers to anticipate fan preferences, predict ticket sales, and identify emerging market opportunities. By leveraging real-time and historical data, sports organizations will have a better understanding of their fans' needs, allowing for personalized marketing campaigns that resonate with individuals on a deeper level. For example, using predictive analytics, a sports team could identify fans who are more likely to purchase merchandise during a particular season or predict which fan segments are most likely to attend specific types of events. This insight can inform targeted marketing efforts, optimize promotional activities, and maximize revenue potential.

Decision intelligence will play a pivotal role in harnessing the power of augmented reality (AR) and virtual reality (VR) technologies to enhance fan experiences. Sports organizations will be able to leverage decision intelligence to create immersive virtual experiences, allowing fans to engage with their favorite teams and athletes from anywhere in the world. By analyzing fan behavior and preferences, marketers can deliver personalized AR/VR content, such as virtual meet-and-greets, behind-the-scenes access, and interactive game experiences, to drive fan engagement and loyalty. For instance, a sports team could use decision intelligence to identify fan segments that are most likely to engage with AR experiences. By offering personalized AR content, such as virtual tours of the team's facilities or interactive training sessions with players, the team can deepen fan connections and create unique experiences that go beyond traditional marketing approaches.

Future trends will continue embracing social media and influencer marketing. Decision intelligence will continue to evolve in its ability to harness the power of social media and influencer marketing. Sports organizations will leverage advanced analytics to identify social media trends, sentiment analysis, and fan engagement metrics to optimize their social media strategies. Decision intelligence will enable sports marketers to identify influential individuals, micro-influencers, and fan advocates who can effectively promote their brand and engage with a wider audience. For example, decision intelligence could be used to identify fans with a significant following on social media platforms who are passionate about a specific team or sport. By partnering with these influential fans, sports organizations can amplify their brand message, increase reach, and drive fan engagement.

Although the integration of the Internet of things (IoT) and wearable technology is present in the sports marketing scientific discourse, this topic will likely grow in popularity. The future of decision intelligence in sports marketing will involve leveraging data from IoT devices and wearable technology. By collecting data from smart stadiums, wearable devices, and connected fan experiences, sports marketers can gain valuable insights into fan behavior, preferences, and engagement levels. This data can be used to optimize marketing campaigns, enhance in-stadium experiences, and provide personalized offers to fans. For instance, wearable devices can provide real-time data on a fan's heart rate, location, and activity levels during a game. Decision intelligence can analyze this data to create personalized offers, such as discounts on concessions or exclusive access to team merchandise, based on a fan's engagement and loyalty

Speaking of future trends, the chapter presents some potential research methodologies for future studies in this area, too. Future researchers can explore the use of decision intelligence techniques in sports marketing using multiple and linear regression analyses, structural equations modeling (SEM), data mining, and similar, which are rooted in ML and AI technologies. Moreover, researchers can employ sentiment analysis, too, to gauge the fans' opinions and feelings. Besides the quantitative approaches, researchers can combine them with qualitative methods including interviews, focus groups, and ethnographic studies to holistically understand the motivations, perceptions, and experiences of fans and sports marketers in their willingness to use decision intelligence techniques.

7 Conclusion

This book chapter has implications for both the academic community and professionals in sports marketing in the sense that they can better understand the current state and the prospects of decision-making, which takes into account as much data as possible, churning and processing it, and coming up with prospective courses of action.

Professional sports marketing and decision-making have a long history of sticking to practices that have previously worked and the so-called gut feeling of marketers. This has changed from the start of the first and second decade of this millennium as statistics at first, and then AI and big data took over. It has been shown that rigorous analytics and statistical analysis through their forecasting techniques represent far superior indicators of success rather than traditional methods.

Nowadays, most professional sports clubs, leagues, and other organizations have analytics experts and even entire teams. So, it is up to them to stay ahead of the competition by finding ways to ethically and responsibly collect more data on their target customers and using state-of-the-art business analytics techniques grounded in machine learning, AI, and sophisticated forecasting methods. Investments in building capacities of analytics teams and all involved parties in sports marketing will more likely pay off as the disruptive nature of the sports environment can only be tamed through comprehensive analytics and clear scenarios. Furthermore, enhanced collaboration within the team and between organizations and stakeholders can act as a boost to the sports marketing efforts, as all newly engaged sports stakeholders can represent a valuable data source.

The examples and applied decision intelligence practices in sports marketing activities outlined in this chapter have several practical implications for business leaders, managers, and employees working in the marketing departments of sports organizations. Considering the presented findings, sports marketers can focus on data-driven strategies and decisions that can help them drive sales, improve the fan experience and social media presence, and optimize sponsorships. By implementing even small-scale decision intelligence interventions in their marketing efforts, these employees can contribute to creating a data-driven decision culture. This involves nurturing an organizational mindset that sports marketers should be educated about the importance of leveraging data and analytics tools to work more effectively and efficiently. By relying on factual data and quantifiable indicators, practitioners can assess the risks and mitigate their potential negative effects on the marketing strategies.

As data gets more diverse so do the data sources, so now analytics teams can analyze fan engagement from fan pages of their organizations on social media and can customize the VIP spectator experience from the food that is bought the most during a game and the most popular seats at a stadium. From data on ticket pricing, sales inventory, and digital marketing analytics to customer relationship management, fan engagement, sponsorship valuation, and market research, there is a wide range of dimensions where predictive analytics can help sports organizations make the right decision.

References

Abdolmaleki, H., Mirzazadeh, Z. S., & Ghahfarokhhi, E. A. (2018). Identify and prioritise factors affecting sports consumer behaviour in Iran. *International Journal of Sport Management and Marketing, 18*(1–2), 42–62.

Albán Bartra, C., Alcántara Sánchez, D., & Sagástegui Cruz, J. (2021). Sports marketing and its relationship in the value creation for the customers of sport center, Trujillo, 2019. In *Artificial Intelligence, Computer and Software Engineering Advances: Proceedings of the CIT 2020 Volume 2* (pp. 29–41). Springer International Publishing.

Apostolou, K., & Tjortjis, C. (2019). Sports analytics algorithms for performance prediction. In *2019 10th International Conference on Information, Intelligence, Systems and Applications (IISA)* (pp. 1–4). IEEE.

Balducci, B., & Marinova, D. (2018). Unstructured data in marketing. *Journal of the Academy of Marketing Science, 46*, 557–590.

Beal, R., Norman, T. J., & Ramchurn, S. D. (2019). Artificial intelligence for team sports: A survey. *The Knowledge Engineering Review, 34*, e28.

Beesetty, Y., Kanhaiya, K., & Vineet, K. (2022). *Artificial intelligence in sports market*. Allied Market Research. Retrieved March 27, 2023, from https://www.alliedmarketresearch.com/artificial-intelligence-in-sports-market-A12905

Bukstein, S. (2016). Evolution and impact of business analytics in sport. In C. K. Harrison & S. Bukstein (Eds.), *Sport business analytics* (pp. 25–46). Auerbach Publications.

Chang, Y. (2019). Spectators' emotional responses in tweets during the super bowl 50 game. *Sport Management Review, 22*(3), 348–362.

Chen, C. Y., Stotlar, D. K., & Lin, Y. H. (2009). Prediction of ticket purchase in professional sport using data mining. *International Journal of Sport Management and Marketing, 6*(1), 68–86.

Cvetkoska, V. (2022). *Business analytics*. Stobi trejd.

da Silva, E. C., & Las Casas, A. L. (2020). Key elements of sports marketing activities for sports events. *International Journal of Business Administration, 11*(1), 11–21.

Dam, N. A. K., Le Dinh, T., & Menvielle, W. (2019). Marketing intelligence from data mining perspective. *International Journal of Innovation, Management and Technology, 10*(5), 184–190.

Davenport, T. (2014). What businesses can learn from sports analytics. *MIT Sloan Management Review*, 10–13.

Dees, W., Walsh, P., McEvoy, C. D., McKelvey, S., Mullin, B. J., Hardy, S., & Sutton, W. A. (2022). *Sport marketing* (5th ed.). Human Kinetics.

Dewi, M. V. K., & Darma, G. S. (2019). The role of marketing & competitive intelligence in industrial revolution 4.0. *Jurnal Manajemen Bisnis, 16*(1), 1–12.

Diehl, M. A., Drayer, J., & Maxcy, J. G. (2016). On the demand for live sport contests: Insights from the secondary market for national football league tickets. *Journal of Sport Management, 30*(1), 82–94.

Drayer, J., Shapiro, S. L., & Lee, S. (2012). Dynamic ticket pricing in sport: An agenda for research and practice. *Sport Marketing Quarterly, 21*(3), 184–194.

Duan, Y., Edwards, J. S., & Dwivedi, Y. K. (2019). Artificial intelligence for decision making in the era of big data–evolution, challenges and research agenda. *International Journal of Information Management, 48*, 63–71.

Egebjerg, N. H., Hedegaard, N., Kuum, G., Mukkamala, R. R., & Vatrapu, R. (2017). Big social data analytics in football: Predicting spectators and TV ratings from Facebook data. In *2017 IEEE International Congress on Big Data (BigData Congress)* (pp. 81–88). IEEE.

Ermakov, A. V., Skarzhynskaya, E. N., & Novoselov, M. A. (2022). Digital transformation of professions in physical education and sport sector. *Theory and Practice of Physical Culture, 3*, 7–9.

Evans, J. R. (2020). *Business analytics: Methods, models, and decisions* (3rd ed.). Pearson.

Fan, S., Lau, R. Y., & Zhao, J. L. (2015). Demystifying big data analytics for business intelligence through the lens of marketing mix. *Big Data Research, 2*(1), 28–32.

Fink-Hafner, D., Dagen, T., Doušak, M., Novak, M., & Hafner-Fink, M. (2019). Delphi method: Strengths and weaknesses. *Advances in Methodology and Statistics, 16*(2), 1–19.

Fullerton, S., & Merz, G. R. (2008). The four domains of sports marketing: A conceptual framework. *Sport Marketing Quarterly, 17*, 90–108.

Gershenfeld, G. (2015). Conjoint analysis for ticket offerings at the Cleveland Indians. *Interfaces, 45*(2), 166–174.

Goebert, C., & Greenhalgh, G. P. (2020). A new reality: Fan perceptions of augmented reality readiness in sport marketing. *Computers in Human Behavior, 106*, 106231.

Guerard, J. B., Jr., & Guerard, J. B. (2013). *Regression analysis and forecasting models* (pp. 19–45). Introduction to Financial Forecasting in Investment Analysis.

Hensley, N. (2022, November 7). Why fans crave predictive analytics-and how sports can deliver them. *Forbes*. Retrieved March 27, 2023, from https://www.forbes.com/sites/forbescommunicationscouncil/2022/11/04/why-fans-crave-predictive-analytics-and-how-sports-can-deliver-them/

Houtmeyers, K. C., Jaspers, A., & Figueiredo, P. (2021). Managing the training process in elite sports: From descriptive to prescriptive data analytics. *International Journal of Sports Physiology and Performance, 16*(11), 1719–1723.

Hubáček, O., Šourek, G., & Železný, F. (2019). Exploiting sports-betting market using machine learning. *International Journal of Forecasting, 35*(2), 783–796.

Jensen, J. A., & Turner, B. A. (2017). Advances in sport sponsorship revenue forecasting: An event history analysis approach. *Sport Marketing Quarterly, 26*(1).

Jensen, J. A., Wakefield, L., Cobbs, J. B., & Turner, B. A. (2016). Forecasting sponsorship costs: Marketing intelligence in the athletic apparel industry. *Marketing Intelligence & Planning, 34*(2), 281–298.

Katsikea, E., Theodosiou, M., & Makri, K. (2019). The interplay between market intelligence activities and sales strategy as drivers of performance in foreign markets. *European Journal of Marketing, 53*(10), 2080–2108.

Lindsey, G. R. (1959). Statistical data useful for the operation of a baseball team. *Operations Research, 7*(2), 197–207.

Mamo, Y., Su, Y., & Andrew, D. P. (2022). The transformative impact of big data applications in sport marketing: Current and future directions. *International Journal of Sports Marketing and Sponsorship, 23*(3), 594–611.

Marr, B. (2015). *Big data: Using SMART big data analytics and metrics to make better decisions and improve performance*. John Wiley & Sons.

Morgulev, E., Azar, O. H., & Lidor, R. (2018). Sports analytics and the big-data era. *International Journal of Data Science and Analytics, 5*, 213–222.

Mumcu, C., & Fried, G. (2017). Analytics in sport marketing. *Sport Management Education Journal, 11*(2), 102–105.

N3XT Sports. (n.d.). *The importance of fan segmentation: how to turn data into engagement*. Retrieved March 28, 2023, from https://www.n3xtsports.com/the-importance-of-fan-segmentation/

Power, D. J., Heavin, C., McDermott, J., & Daly, M. (2018). Defining business analytics: An empirical approach. *Journal of Business Analytics, 1*(1), 40–53.

Prasetio, D., & Harlili, D. (2016, August). Predicting football match results with logistic regression. In *2016 International Conference On Advanced Informatics: Concepts, Theory and Application (ICAICTA)* (pp. 1–5). IEEE.

Putka, D. J., Beatty, A. S., & Reeder, M. C. (2018). Modern prediction methods: New perspectives on a common problem. *Organizational Research Methods, 21*(3), 689–732.

Quarterman, J., Pitts, B. G., Jackson, E. N., Kim, K., & Kim, J. (2005). Statistical data analysis techniques employed in the sport marketing quarterly 1992 to 2002. *Sport Marketing Quarterly, 14*(4), 227–238.

Ranjan, J., & Foropon, C. (2021). Big data analytics in building the competitive intelligence of organizations. *International Journal of Information Management, 56*, 102231.

Ransbotham, S., Kiron, D., & Prentice, P. K. (2016). Beyond the hype: The hard work behind analytics success. *MIT Sloan Management Review, 57*(3).

Sedky, D., Kortam, W., & Abou Aish, E. (2022). The role of sports marketing in attracting audiences towards less popular sports. *Journal of Humanities and Applied Social Sciences, 4*(2), 113–131.

Snyder, K., McKelvey, S., & Sutton, W. (2016). All together now? Exploring sales and marketing integration. *Sport, Business and Management: An International Journal, 6*(1), 2–18.

Stein, G. (2014). Will ticket scalpers meet the same fate as spinal tap drummers? The sale and resale of concert and sport tickets. *Pepperdine Law Review, 42*, 1–52.

Stein, F., & Greenland, A. (2014). Producing insights from information through analytics. In J. Liebowitz (Ed.), *Business analytics: An introduction* (pp. 29–54). Taylor & Francis Group, LLC/CRC Press.

Su, Y., Baker, B. J., Doyle, J. P., & Yan, M. (2020). Fan engagement in 15 seconds: Athletes' relationship marketing during a pandemic via TikTok. *International Journal of Sport Communication, 13*(3), 436–446.

Tonidandel, S., King, E. B., & Cortina, J. M. (2018). Big data methods: Leveraging modern data analytic techniques to build organizational science. *Organizational Research Methods, 21*(3), 525–547.

Troilo, M., Bouchet, A., Urban, T. L., & Sutton, W. A. (2016). Perception, reality, and the adoption of business analytics: Evidence from North American professional sport organizations. *Omega, 59*, 72–83.

Vale, L., & Fernandes, T. (2018). Social media and sports: Driving fan engagement with football clubs on Facebook. *Journal of Strategic Marketing, 26*(1), 37–55.

Venter, P., & Jansen van Rensburg, M. (2014). The relationship between marketing intelligence and strategic marketing. *South African Journal of Economic and Management Sciences, 17*(4), 440–456.

Watanabe, N. M., Shapiro, S., & Drayer, J. (2021). Big data and analytics in sport management. *Journal of Sport Management, 35*(3), 197–202.

Wedel, M., & Kannan, P. K. (2016). Marketing analytics for data-rich environments. *Journal of Marketing, 80*(6), 97–121.

Yarkoni, T., & Westfall, J. (2017). Choosing prediction over explanation in psychology: Lessons from machine learning. *Perspectives on Psychological Science, 12*(6), 1100–1122.

Yeoh, W., & Popovič, A. (2016). Extending the understanding of critical success factors for implementing business intelligence systems. *Journal of the Association for Information Science and Technology, 67*(1), 134–147.

Zhang, H., & Zhu, J. (2022). Practicability of sports goods in the sports field based on artificial intelligence technology. *Mobile Information Systems*, 2022.

Digital Marketing Analytics in Sports

Ali B. Mahmoud

1 Introduction

This chapter peers into digital marketing analytics in sports. It deals with how data and analytics are changing marketing in this field. Digital marketing analytics uses technology to collect and make sense of data. It uses digital platforms like social media, sites, and mobile apps. This data, which spans fan actions to ads' performance, provides key info for sports companies (Chaffey & Smith, 2022; Davenport & Harris, 2007; Eagleman, 2013; Kramer et al., 2021; Mamo et al., 2022; Pappalardo et al., 2021; Read & Smith, 2023; Wood et al., 2019; Yadav et al., 2023).

Digital marketing analytics in sports has huge importance. In an age where digital media dominates, sports companies are discovering new ways to connect with fans and add new income sources. For example, the Golden State Warriors are leaders in merging data analytics and fan connection strategies. As Boulton (2018) observes, they're customizing fan encounters, from making purchases at games simpler to providing unique digital events like the "Dub Hub" virtual experience (Murray, 2021). These approaches are coupled with smart mobile, digital, and social marketing tactics, demonstrating the team's commitment to using tech and data insights to reach fans (Johnson, 2017).

This chapter delivers an in-depth exploration of digital marketing analytics in sports. This includes its use in fan engagement, sponsorship, advertising, ticketing, and boosting revenues. This is set up for practitioners, teachers, and students interested in decoding this budding domain. It combines scholarly ideas with real-world samples. We will feature current trends and hurdles in the field, illustrating how

A. B. Mahmoud (✉)
St. John's University, New York, USA

University of London, London, UK
e-mail: mahmouda@stjohns.edu

A. Mansurali et al. (eds.), *Sports Analytics*,
https://doi.org/10.1007/978-3-031-63573-1_4

sports entities can use digital marketing analytics to stay on top in the rapidly advancing digital world.

2 The Importance of Marketing Analytics in Sports

The fusion of marketing analytics with the sports sector signifies a monumental shift for both scholarly research and the sports business landscape. In the modern era, sports transcend the realm of mere athletic contests; they embody sophisticated business entities that necessitate a profound grasp of factors such as fan engagement, the valuation of sponsorships, and consumer dynamics (Carrillat & d'Astous, 2012). Utilizing a data-centric approach, sports marketing analytics leverages tools like predictive modeling and customer segmentation to enhance ticket sales tactics, customize experiences for fans, design targeted marketing initiatives, and maximize the efficacy of sponsorship agreements (Kim et al., 2019). This approach not only propels financial outcomes to new heights but also equips decision-makers with the insights needed to navigate an ever-shifting market, ensuring strategies remain attuned to the latest trends and audience preferences (Jin, 2017).

2.1 Digital Marketing Analytics for Fan Engagement

At the heart of effective sports marketing is a deep understanding of fan behavior and preferences. Digital marketing analytics empower sports organizations to collect multifaceted data on fans' viewing patterns, social media activities, and purchasing habits. Tools like sentiment analysis, as detailed in previous research (e.g. Mahmoud, 2024a, b; Mahmoud et al., 2024), help teams gauge the emotional landscape surrounding games, players, and marketing campaigns. For instance, IBM has used its AI capabilities to enhance the ESPN Fantasy Football experience by analyzing vast quantities of structured and unstructured data to help players understand the value of potential trades and make sense of matchups (IBM, 2022). Additionally, IBM has used its Watson Explorer tool to analyze the organic conversation and pick up on sentiment in the context of football (Balduino, 2018).

Segmentation techniques enable the identification of specific fan groups, each displaying unique characteristics and requirements. Leveraging clustering algorithms and data mining methods, teams can categorize fans by engagement metrics and buying tendencies (Oyewole & Thopil, 2023; van Dam & van de Velden, 2015). One example is the use of data and AI by the National Football League (NFL) to improve ticket sales and fan engagement. In 2021, Bobby Gallo, Senior Vice President of Club Business Development at the NFL, presented a slide to NFL team owners showing teams with at least 15,000 unsold tickets on average for the upcoming season (Wittenberg & Meaney, 2022).

Similarly, the value of analytics extends far beyond mere metrics in social media, turning it into a dynamic platform for fan interaction. Such techniques, which include sentiment analysis and trend tracking, turn fan insights into actionable strategies, elevating the brand experience (Di Virgilio et al., 2022). Furthermore, often overlooked direct channels like email marketing transform into powerful instruments for fan outreach when infused with analytics. For example, the use of CRM by the National Football League team used to help gauge individual customers' ability to pay for season tickets, ensuring the team's sales force had the right data when approaching a potential season ticket holder (Scott, 2018).

Subsequently, the advent of advanced digital marketing analytics revolutionizes personalization in sports marketing. Real-time data serves as an evolving narrative of fan engagement, illustrated by how Nike is using analytics to personalize their customer experience. Nike has developed an app to enhance its interaction with customers. This app allows customers to access the Nike Plus rewards program, which provides exclusive and early access to new products. One of the app's unique features is access to Nike experts for any sport you play. Additionally, Nike offers personalized workouts and priority access to its events (Barseghian, 2019). Emerging technologies like machine learning and AI act as enablers, making marketing strategies agile and exceptionally responsive to each fan's unique preferences (Mohammadi et al., 2023). That said, Marketing analytics equips sports organizations with the power to understand their audience through a data-driven lens. This strategic approach enhances fan engagement, strengthens brand loyalty, and drives revenue growth.

2.2 *Digital Marketing Analytics for Sponsorship and Advertising*

In the competitive arena of the sports industry, digital marketing analytics is an essential asset. It sheds light on the effectiveness of sponsorships and advertising campaigns by analyzing key performance indicators. These include brand exposure, engagement rates, and conversion metrics, as noted by Chmait et al. (2020), Green (2018), and Maricic et al. (2019). A notable example is Coca-Cola's initiative to enhance biodiversity around the Olympic venues in PyeongChang through a dedicated water replenishment program. Furthermore, since the London 2012 Olympic Games, the company has committed to sustainability by offering its products in 100% recyclable plastic bottles, according to olympics.com (2019). The use of real-time analytics to gauge brand sentiment and visibility across digital channels can help Coca-Cola with immediate adjustments to enhance the effectiveness of sponsorships and advertising.

Monitoring digital media performance is vital for understanding how advertising campaigns and sponsorships resonate with audiences (Jarvis & Westcott, 2021; Pykes, 2022; Schroer, 2022; Vitale, 2021). Tools like Google Analytics, Adobe

Analytics, and other proprietary platforms enable sports organizations to track user behaviors, click-through rates, time spent on advertisements, and other critical metrics. One such organization is Sportradar, which helps sports leagues, such as the K-League in Korea, with their platform strategy, including whether to charge for content (Lim, 2023). Sportradar also recently launched a new platform for baseball in Australia, working with the Australian Baseball League (Lim, 2023).

Analyzing the return on investment (ROI) for sponsorships and advertising ensures that sports organizations maximize the impact of their marketing budget, which involves assessing the tangible and intangible returns relative to the costs involved (Jensen, 2023). Mercedes-Benz conducted a comprehensive ROI analysis for their sponsorship of the Mercedes-Benz Stadium in Atlanta. According to Tucker (2016), the Falcons organization sold corporate sponsorships at Mercedes-Benz Stadium, totaling up to $900 million in contractually obligated long-term revenue, putting them within $100 million of their $1 billion contracted sponsorship revenue goal. Moreover, there were 29 total activations and 18 sponsored bars, with some impressive activations, including the AT&T Perch, American Family Insurance Fan Village, and the Home Depot Backyard (Wood & Smith, 2018).

Digital marketing analytics is more than a tracking tool; it is a strategic instrument to optimize sponsorship and advertising strategies. Through predictive modeling and A/B testing, sports organizations can forecast the likely success of various sponsorship and advertising approaches (Lough & Geurin, 2019). Adidas has been known to use machine learning algorithms and AI to boost its personalized email marketing game. They have used data points like language, location, and product interest to match subscribers with the most relevant content (Peymandoust, 2021). Additionally, Adidas has also used Google data insights to create targeted storytelling and more relevant product messages (Digital Strategy Consulting, 2020).

2.3 *Digital Marketing Analytics for Ticketing and Revenue*

Digital marketing analytics provides an essential framework for understanding ticket sales patterns and trends within the sports industry. Analyzing historical data, such as seasonality, opponent popularity, and game significance, can reveal key insights for ticketing strategies (Quansah et al., 2023). Yankees use data analytics to study past ticket sales and identify optimal pricing and positioning for various seating sections; it is reasonable to assume that a team with such a strong focus on analytics would apply these techniques to all aspects of their operations, including ticket sales. As reported by LaLoggia (2019), the New York Yankees have significantly ramped up their investment in analytics, resulting in the formation of a robust and dynamic analytics department within the past ten years. The team's first analyst, Michael Fishman, was hired in 2005 and helped create a quantitative analysis team that now employs over 20 analysts (LaLoggia, 2019). Fishman also changed the scouting department by creating a database incorporating video analytics and qualitative analysis from scouts (LaLoggia, 2019). This initiative eventually resulted in their proprietary analytics system, the Baseball Analysis and Statistics Engine (BASE).

Predictive modeling has become a powerful tool in forecasting ticket sales. Sports organizations can create strong sales models by analyzing historical data, team performance, weather, and market demographics (Dees et al., 2021; Miller, 2015; Smith & Stewart, 2010). For instance, Event Dynamic uses machine learning to optimize attendance rates instead of revenue maximization in sports ticketing. Using artificial intelligence, they have significantly reduced the manual labor involved in ticket pricing and can accurately forecast what fans are willing to pay. This approach has been notably successful; for example, they collaborated with a collegiate football team and achieved a season record for ticket revenue by aligning prices with what would draw a larger crowd of moderate fans (Kern, 2020). Additionally, the company has adapted to the changing landscape with its Stadium Configurator, a feature that lets sports organizations customize seating charts in real-time to meet social distancing regulations (Kern, 2020). The system continuously updates based on ticket sales, ensuring that fans from different parties are not seated next to each other, thus maximizing operational efficiency while prioritizing fan safety and experience (Kern, 2020).

The use of analytics in sports organizations extends beyond ticketing to revenue optimization in merchandising, concessions, sponsorships, and other revenue streams (Beatriz & Santos, 2022; Yılmaz et al., 2023). One sports company that uses analytics to optimize revenue across multiple channels, including targeted marketing efforts that drive merchandise sales and in-stadium spending, is the Dallas Cowboys (Becker et al., 2021). They use a data platform called Legends that collects and analyzes data from various sources, such as ticketing, concessions, merchandise, social media, and fan surveys (Becker et al., 2021). Such platforms and tools can help those organizations understand their fans' preferences, behaviors, and spending patterns and tailor their offers and promotions accordingly.

Finally, using analytics in the sports industry has many benefits for revenue optimization, ticket pricing, and predictive modeling. For instance, a study by Boldt et al. (2016) used Facebook data and event studies to successfully forecast sales trends for Nike. The study applied computational social science methods to gauge the informational value of social media metrics, hence informing marketing strategies and sales operations. Another example relates to the NBA's embracing of data analytics in a way that surpasses most other major US sports leagues (McLaughlin, 2018). They use data analytics tools to improve their decision-making, which enables them to plan better and innovate faster (McLaughlin, 2018). Also, a study by Chmait and Westerbeek (2021) used machine learning to forecast the demand and sales of tickets for Australian football league (AFL) matches. They applied linear regression, random forest, and neural network models to historical data and found that the neural network model achieved the highest accuracy (Chmait & Westerbeek, 2021). Yet, while analytics extends beyond ticket sales to include additional revenue streams (e.g., merchandise and concessions), an investigation by Bouchet et al. (2016) that surveyed and interviewed high-ranking executives and managers in North American sports organizations on the use of dynamic pricing in

North American professional sports found that while most organizations perceive high usage of dynamic pricing within their organization, current procedures lag suggesting that there is still ample room for improvement. Overall, analytics is a powerful tool for sports organizations to craft precise and effective revenue strategies, from predicting ticket sales to optimizing pricing and expanding income channels.

3 Sources and Methods

Data collection in sports marketing analytics has evolved with technological advancements. Various methods are employed, ranging from online surveys to sophisticated tools like web scraping and wearable technologies (Giusti Gestri, 2023; Naraine et al., 2022; Yadav et al., 2022; Zhu et al., 2023). Social media platforms offer a rich source of fan interaction and sentiment data, while ticketing systems provide insights into consumer purchasing behaviors. Wearable devices and sensors, used by many professional teams, provide real-time performance data that can be utilized for fan engagement (Mathew & George, 2022; Zagnoli & Radicchi, 2022).

3.1 Types of Data Collected

Sports marketing analytics involves collecting diverse data types to create targeted strategies that connect with fans and stakeholders. This multifaceted approach enriches the landscape of data, covering various dimensions.

Fan data plays a pivotal role in understanding the pulse of the audience. This includes preferences, behaviors, and interactions on social media, providing insights that shape marketing efforts (Schwarz et al., 2022). For instance, Manchester United's digital fan engagement strategy during the pandemic boosted its commercial revenue and online offerings (StriveCloud, 2022). The club also appointed its first Head of Fan Engagement to better connect with fans and take their interests into account when making decisions (Man Utd Communications Department, 2022). Furthermore, Arsenal FC developed a data-driven fan engagement app that provided personalized content and real-time match statistics to its fans (Adair, 2022). These examples show how fan data can help sports organizations to enhance the fan experience and loyalty.

Ticketing data offers another rich source, revealing patterns in attendance and spending. Major League Baseball (MLB), for example, has analyzed this data to optimize pricing and promotions (Rascher et al., 2007). Alongside this, ***sponsorship and advertising data*** form a vital component, enabling tracking of the effectiveness of marketing campaigns and aligning strategies with commercial goals (Fullerton, 2021). However, measuring sponsorship effectiveness can be a complex task, particularly when contrasted with traditional advertising measurements. According to

B2B International (2023), around $600 billion is spent on advertising annually, with sponsorship investments exceeding $50 billion. While advertising has refined methods for gauging impact, sponsorship's effectiveness remains more elusive, often involving branding on apparel, vehicles, or sports venues to enhance brand awareness and shape perceptions. Bennett (2021) stresses the need to evaluate marketing effectiveness by concentrating on metrics such as revenue, ROI, conversion rates, and retention rates, ensuring optimal return on marketing investment (ROMI). Giorgio (2017) adds a perspective from sports, showing that comprehensive data about fans' behaviors and preferences can enable more targeted and authentic sponsor engagements both inside and outside the stadium. Finally, Lyons (2023) highlights the opportunity sponsorships present to increase brand visibility, noting that these connections can lead to extensive exposure.

Performance data, including player statistics, is vital in building a holistic picture of the sporting ecosystem. The advent of cutting-edge technology, like the NBA's use of SportVU cameras, captures detailed player movements and has revolutionized how teams analyze performance (Westney, 2013). Clubs like FC Barcelona have further harnessed wearable technology and biometrics to craft interactive narratives that enhance fan engagement. In August 2022, FC Barcelona announced a strategic partnership with Socios.com, a blockchain-powered fan engagement and rewards platform. The partnership aims to accelerate FC Barcelona's blockchain, NFT, and digital assets strategies (Socios., 2022). Additionally, in October 2018, Viber integrated consumer activation platform Vivoom into its FC Barcelona chatbot, enabling visitors to Camp Nou to create and share unique end-to-end videos (Srivastav, 2018).

The rise of ***social media*** has opened additional avenues for engagement analysis (López-Carril et al., 2020; McCarthy et al., 2022; Naraine et al., 2019; Shabbirhusain et al., 2023), thus redefining communication within sports fan communities, creating virtual spaces that impact both bonding and bridging aspects of social capital (Fenton et al., 2021; McCarthy et al., 2022). Following this, organizations like the NBA monitor fan interactions across platforms to inform content creation (Moran, 2020; Sim, 2023). Similarly, in the broadcast realm, analysis of television and streaming viewing patterns allows leagues like the English Premier League to schedule content that aligns perfectly with global audiences (Hutchins et al., 2019). In this vein, ethnographic and netnographic research methods offer deep, qualitative insights into the cultural and behavioral aspects of sports fans, both offline and online. For instance, ethnographic studies might involve researchers immersing themselves in fan communities. Netnography, on the other hand, focuses on analyzing online platforms like fan forums or social media to understand virtual behaviors, similar to studies conducted on eSports communities. These methods allow sports marketers to uncover complex social dynamics, emerging trends, and fan sentiments that quantitative data alone cannot capture. Thus, they serve as invaluable tools for crafting more effective and targeted sports marketing strategies (Giulianotti, 2015; Kozinets, 2019) and using netnography, aka digital ethnography, as a research method allows researchers to observe how people interact in online spaces unobtrusively (Armstrong et al., 2014; Kozinets, 2002; Murthy, 2011). This means that by conducting digital ethnography, it is possible to collect and analyze information and interactions that happened in cyberspace. Clubs like Salford City Football Club (SCFC) have leveraged these social media brand communities (SMBCs) to support local initiatives and build a sense of place, which is especially important in the era of COVID-19-induced restrictions (Fenton et al., 2021).

3.2 Data Analysis Techniques Used in Sports

Data analysis in sports marketing requires a blend of traditional statistical methods and modern machine-learning techniques (Bai & Bai, 2021; Chmait & Westerbeek, 2021). *Descriptive analytics* provide a snapshot of historical data, while predictive analytics forecast future trends, such as expected ticket sales. Advanced techniques like *sentiment analysis* (Mahmoud et al., 2021) are used to gauge fan emotions on social media, and network analysis maps the interconnectivity within the sports community (O'Hallarn et al., 2019). *Clustering algorithms*, a specialized machine learning technique (Anifa et al., 2022), offer significant utility in sports marketing. For example, FC Barcelona's real-life utilizes clustering to effectively segment and target different fan groups (FC Barcelona, 2020). In the diverse world of sports analytics, we encounter a variety of algorithm types, each serving a distinct purpose. From partitioning and hierarchical to density-based, grid-based, and model-based algorithms (Anifa et al., 2022), these methods use mathematical principles to systematically categorize data into various groups. Consider the case of *FC Barcelona*, a club that masterfully employs clustering algorithms to probe the depths of its extensive fan base. This sophisticated analysis empowers the club to design marketing strategies that resonate deeply with the diverse tastes and preferences of its fans, enhancing their engagement and satisfaction. Yet, the effectiveness of these strategies hinges on the presence of meticulously prepared data and robust automated data management and reporting systems. These foundational elements are crucial in leveraging the vast capabilities of analytical algorithms. Our discussion highlights the various methodologies that are pivotal in the realm of sports marketing analytics, each serving specific strategic purposes.

3.2.1 Fan Engagement and Segmentation

In the competitive landscape of the sports industry, the ability to captivate fans and intelligently segment them is essential for sustaining and expanding both revenue and brand devotion. Clustering algorithms stand out as transformative tools; they dissect engagement metrics, allowing teams to categorize their fans—from the die-hard enthusiasts to the casual spectators. Such dynamic segmentation paves the way for marketing initiatives that hit the mark (de Vries et al., 2019; Venkatesan et al., 2021).

Additionally, sentiment analysis has become indispensable in gauging fan sentiment on social media, offering real-time insights that can steer strategic moves ranging from PR tactics to customer service (Liu et al., 2022). Accurate fan sentiment analysis is vital for crafting communications and promotions that align with the fans' prevailing emotions and preferences.

In a proactive sense, *Churn Prediction Models* are invaluable for spotting fans who may be on the brink of disengagement. These models, which scrutinize past interactions and behavioral trends, inform retention strategies that are key to preserving a loyal fan base. Collectively, these analytical methods provide a holistic view of the fan landscape, a necessity in the dynamic domain of sports marketing.

3.2.2 Ticket Sales and Pricing

The field of ticket sales and pricing has been transformed by strategic data analytics. *Predictive modeling*, with its capacity to forecast sales trends by drawing from historical data, grants sports organizations the vision needed to optimize their inventory and refine their marketing tactics (Begum, 2023; Kennedy et al., 2021). This strategic foresight empowers teams to customize their budgets thoughtfully, enabling adjustments in advertising investments and promotional strategies to be perfectly synchronized with anticipated sales trends. At the same time, *price elasticity models* delve into the dynamics between pricing and consumer demand, offering priceless insights. These insights reveal how variations in ticket pricing could influence sales volumes (Yılmaz et al., 2023). Armed with this knowledge, organizations can implement versatile pricing frameworks, adeptly capturing a diverse range of consumer segments and maximizing revenue potential. This allows organizations to adopt flexible pricing structures to capture various consumer segments effectively. Finally, *dynamic pricing* approaches add another layer of sophistication by enabling real-time price adjustments based on real-time demand metrics, competitor pricing, and other market dynamics (Yılmaz et al., 2023). Sports organizations have the ability to significantly increase their revenue and boost event attendance by employing data-driven analytical methods. These strategies are adept at crafting highly effective ticket sales campaigns that resonate with audiences.

3.2.3 Social Media Analytics

In the realm of sports marketing, harnessing the power of social media analytics is vital for crafting compelling narratives and forging deeper connections with fans. By utilizing *time-series analysis*, marketers can gain invaluable insights into how fan engagement and the performance of content evolve over time (Yan et al., 2023).

This helps teams pinpoint optimal posting times, identify seasonality in fan interactions, and adjust their content calendar accordingly. *Social monitoring and listening* tools further amplify the ability to proactively manage brand image by monitoring real-time mentions, trends, and public sentiment related to the team or specific events (Zachlod et al., 2022). Such insights can be particularly valuable for crisis management or leveraging viral moments. Additionally, *competitive analysis* provides a benchmark to assess the efficacy of a team's social media strategy by comparing key performance indicators like engagement rates, follower growth, and content virality with industry competitors (Grant, 2021; Šerić & Ljubica, 2018). By utilizing these analytics techniques, sports organizations can fully optimize their social media approaches, resulting in enhanced fan engagement and increased opportunities for revenue growth.

3.2.4 Merchandising

Merchandising is a crucial revenue stream for sports organizations, and leveraging analytics can significantly refine this aspect of operations. *Inventory forecasting* methods employ historical sales data and current market trends to predict the future demand for merchandise items, thus helping teams to balance stock levels and minimize carrying costs effectively (Schwarz & Hunter, 2017). This proactive approach reduces the risk of overstocking or understocking, which can directly impact profitability. Further enhancing the merchandising strategy is *RFM (recency, frequency, money) analysis*, which categorizes customers based on their purchasing behavior (Valendin et al., 2022). Sports organizations can identify their most valuable customers for targeted marketing initiatives by analyzing the recency, frequency, and monetary value of their transactions (Rogic & Kascelan, 2019). These high-value fans are more likely to respond to premium product offerings or loyalty programs. Together, inventory forecasting and RFM analysis contribute to a more strategic, customer-centric approach to merchandising in the sports industry.

3.2.5 Game-Day Revenue

Game-day revenue represents a critical component of a sports organization's financial health, and sophisticated analytics are increasingly being employed to maximize this income stream (Harrison & Bukstein, 2022). *Concession sales analysis* uses data-driven approaches to determine the optimal types and quantities of food and beverages to stock, as well as the ideal mix of merchandise to offer (Goff Inglis & Zolfaghari, 2017; Mumcu, 2019). By analyzing historical sales data, seasonal trends, and even weather forecasts, teams can better anticipate fan preferences and adjust their concession strategies accordingly. Concurrently, *seat map analysis* leverages spatial analytics to assess the most profitable venue sections based on ticket sales, attendance, and additional in-seat purchases (Courty & Davey, 2019; Glebova et al., 2023). This granular level of insight informs dynamic pricing strategies and targeted promotional campaigns, allowing organizations to fill seats in less popular sections and optimize pricing for high-demand areas. Together, *concession sales analysis* and *seat map analysis* provide sports organizations with actionable intelligence to enhance game-day revenue streams in a manner that is both efficient and responsive to fan preferences.

3.2.6 Player Analytics

Within the dynamic realm of the sports industry, *player analytics* assume a pivotal role in both elevating team performance and driving commercial achievements to new heights. *Player valuation models* provide a suite of quantifiable metrics that adeptly measure a player's influence on key business outcomes, including but not limited to, robust ticket sales, soaring viewership ratings, and substantial merchandise revenue. Cited in Chmait et al. (2020), these models incorporate complex algorithms to meticulously evaluate aspects such as a player's on-field prowess, market appeal, and the breadth of their fan base. The insights gleaned empower teams with data-driven clarity for crafting astute investment strategies and shrewd contract negotiations.

Building upon this foundation is the layer of *player performance metrics*. As per the insights of Aroganam et al. (2019), this facet entails the meticulous, real-time analysis of athletes' skills and capabilities, frequently augmented by state-of-the-art wearable tech and advanced biometric systems. By deftly correlating these performance indicators with levels of fan engagement, sports entities can pinpoint the athletes who truly resonate with and enthrall their audiences. This intelligence becomes a linchpin for conceiving focused marketing initiatives and strategically showcasing star players at the zenith of pivotal matches or throughout a riveting season.

The synergistic amalgamation of *player valuation models* and *player performance metrics* furnishes sports organizations with an exceptionally potent analytical arsenal, poised to revolutionize the industry. This dual approach doesn't just enhance the athletic performance on the field; it also sharpens the commercial strategies off the field. By understanding both the athletic prowess of players and their impact on fans, organizations can make informed decisions that benefit the entire operation, from gameplay to fan engagement and revenue generation. This integrated analysis is vital for modern sports teams striving to excel in both competitive play and business success.

3.2.7 Digital and Email Marketing

In today's sports industry, digital and email marketing have become essential channels for engaging fans and generating revenue. The role of analytics in refining these methods is crucial. *A/B testing*, for instance, is a popular technique used to enhance email campaigns. By experimenting with different subject lines, content formats, and calls to action, organizations aim to boost click-through rates and engagement (Krüger, 2023). These experiments offer valuable insights into consumer preferences, enabling more precise and impactful messaging.

On the digital front, the significance of *search engine optimization (SEO)* and *search engine marketing* (SEM) analytics cannot be overstated (Singh, 2023). Monitoring metrics like click-through rates, impressions, and conversion rates helps sports organizations fine-tune their online content and advertising strategies for better search engine performance (Doyle, 2016). By integrating strategies that draw more organic traffic, our approach significantly boosts the return on investment for digital campaigns. Our digital marketing toolkit, which includes *A/B testing* for email campaigns alongside SEO/SEM analytics, provides a comprehensive solution that enhances both fan engagement and revenue streams.

3.2.8 Revenue Optimization

Mastering revenue optimization in the sports sector demands an in-depth analysis of customer behavior and the effectiveness of marketing channels. At the heart of this endeavor lies *lifetime value modeling*, a technique for estimating the long-term worth of a customer. This methodology takes into consideration various elements such as purchase history, engagement levels, and anticipated future interactions, as referenced by Padmanabhan and Tuzhilin (2003). Employing this model, sports organizations are empowered to distribute their resources with a focus on nurturing high-value customers over an extended period.

Moreover, the role of *cross-channel analytics* cannot be understated. It is essential for understanding the particulars of customer interactions across diverse platforms and channels. It provides a comprehensive perspective of how different marketing channels, such as social media, email, and in-game promotions, work together. This insight allows for the development of more cohesive and synergistic marketing strategies, enhancing the overall effectiveness of marketing efforts (Mullin, 2018). Put it forward, revenue optimization in sports is about strategically combining various analytical tools to understand and cater to customer needs effectively, ensuring sustained growth and profitability. These analytics can identify the best-performing channels and help in reallocating marketing spend accordingly. To complete this robust analytical framework, *multi-touch attribution models* are utilized to trace and credit each touchpoint in the customer journey, whether it's an initial social media ad or a final in-app purchase, for its role in conversion (Pattanayak et al., 2022). This approach to revenue optimization balances short-term gains with long-term customer value by ensuring adequate attribution of all marketing efforts.

3.2.9 Real-Time Analytics and Personalization

In the sports industry, the advent of real-time analytics and personalization technologies has fundamentally transformed fan engagement and marketing strategies. *Real-time data analysis* allows sports organizations to capture and interpret data instantly during games or events, enabling immediate insights into fan behavior, game dynamics, and even merchandise sales (Mamo et al., 2022). These insights can be swiftly translated into actionable adjustments, such as tweaking promotional offers or enhancing in-game entertainment. *AI and machine learning algorithms* augment this capability, facilitating advanced predictive analytics and driving personalized marketing initiatives. These algorithms can sift through massive datasets to predict future fan behavior and can adapt marketing campaigns on the fly to resonate with individual fans based on their past interactions and preferences (Pradeep et al., 2018). The synergy between real-time data and advanced algorithms offers sports organizations an unprecedented level of agility and customization in their marketing and operational efforts, thereby creating more engaging and lucrative fan experiences.

4 Future Trends and Challenges

Emerging technologies are continuously evolving the digital marketing analytics landscape in sports, offering both remarkable opportunities and posing significant challenges. Artificial intelligence (AI) and machine learning (ML) are revolutionizing areas like predictive modeling and fan engagement. For instance, the NFL uses AI algorithms to tailor marketing strategies based on fan sentiment (Wittenberg & Meaney, 2022). Blockchain technology, notably implemented by the Sacramento Kings for ticketing, adds a layer of security and transparency to transactions (Kings Rocket Department, 2021). Virtual and augmented reality (VR and AR) are enriching fan experiences, exemplified by the Golden State Warriors' virtual stadium tours (NBA, 2020). Additionally, the Internet of things (IoT) is enabling smart stadiums and connected fan experiences, as seen in FC Barcelona's Camp Nou stadium (The Stadium Business, 2020).

However, these advancements come with their set of challenges and limitations. Data privacy and security are primary concerns, especially with stringent regulations like the General Data Protection Regulation (GDPR). The integration of diverse data sources poses compatibility issues, requiring robust systems for seamless operation. The sports sector is currently grappling with a significant skill gap, especially in the area of data analytics. As the field continues to evolve, the demand for professionals skilled in this domain is on the rise. Alongside this, ethical considerations, particularly around the responsible collection and use of fan data, are becoming increasingly important.

5 Conclusion

Throughout this chapter, we've taken a deep dive into the world of digital marketing analytics in sports. We've seen how integral these analytics are in grasping fan engagement, refining ticket sales and revenue strategies, and evaluating the effectiveness of sponsorship and advertising endeavors. The landscape is constantly shifting, driven by advancements in AI, machine learning, and blockchain technology. These developments present both challenges and opportunities. Sports organizations must stay agile, focus on data-driven strategies, and adhere to ethical standards to fully leverage the benefits of analytics.

The chapter has also underscored key trends and hurdles, highlighting the importance of an interdisciplinary approach, robust data collection, and compliance with data privacy regulations. Looking to the future, it's clear that sports organizations must not only adopt new technologies but also develop flexible marketing strategies that are mindful of ethical and regulatory concerns, especially in the realm of data privacy and security.

As we move forward, it's crucial for both practitioners and researchers in the field to collaborate and continuously reevaluate their strategies. The ultimate aim should be to craft fan-centric experiences that consider the diverse needs of global markets while adhering to ethical standards. By doing so, the sports industry can ensure a more engaging, effective, and responsible approach to digital marketing in the years to come.

References

Adair, M. (2022). *Arsenal reveal how they use data to boost fan engagement.* Bdaily Ltd. Retrieved 21 August from https://bdaily.co.uk/articles/2022/05/12/arsenal-reveal-how-they-use-data-to-boost-fan-engagement

Anifa, M.,. P.,. M. J., Hack-Polay, D., Mahmoud, A. B., & Grigoriou, N. (2022). Segmenting the retail customers: A multi-model approach of clustering in machine learning. In P. Keikhosrokiani (Ed.), *Handbook of research on consumer behavior change and data analytics in the socio-digital era* (pp. 25–50). IGI Global. https://doi.org/10.4018/978-1-6684-4168-8.ch002

Armstrong, C. G., Delia, E. B., & Giardina, M. D. (2014). Embracing the social in social media: An analysis of the social media marketing strategies of the Los Angeles Kings. *Communication & Sport, 4*(2), 145–165. https://doi.org/10.1177/2167479514532914

Aroganam, G., Manivannan, N., & Harrison, D. (2019). Review on wearable technology sensors used in consumer sport applications. *Sensors*. https://doi.org/10.3390/s19091983

B2B International. (2023). *How to measure sponsorship effectiveness and ROI.* B2B International. Retrieved 24 August from https://www.b2binternational.com/publications/sponsorship-research/

Bai, Z., & Bai, X. (2021). Sports big data: Management, analysis, applications, and challenges. *Complexity, 2021*, 6676297. https://doi.org/10.1155/2021/6676297

Balduino, R. (2018). *Football fan sentiment.* IBM. Retrieved 23 August from https://community.ibm.com/community/user/ai-datascience/blogs/ricardo-balduino/2018/10/22/football-fan-sentiment

Barseghian, A. (2019). How nike is using analytics to personalize their customer experience. *Forbes*. Retrieved 27 August from https://www.forbes.com/sites/forbestechcouncil/2019/10/07/how-nike-is-using-analytics-to-personalize-their-customer-experience/?sh=7c6ff30d1611

Beatriz, M. F., & Santos, V. (2022). Stadium 2.0: Framework to improve sports fans' experience in stadium through IoT technology. In T. Guarda, S. Anwar, M. Leon, & F. J. Mota Pinto (Eds.), *Information and knowledge in internet of things* (pp. 229–247). Springer International Publishing. https://doi.org/10.1007/978-3-030-75123-4_11

Becker, S., Berg, A., Kohli, S., & Thiel, A. (2021). *Sporting goods 2021: The next normal for an industry in flux.* McKinsey & Company. Retrieved 28 August from https://www.mckinsey.com/~/media/mckinsey/industries/retail/our%20insights/sporting%20goods%202021%20the%20next%20normal%20for%20an%20industry%20in%20flux/sporting-goods-2021-report.pdf

Begum, T. (2023). Predictive analytics for machine learning and deep learning. In *Handbook of big data research methods* (pp. 148–164). Edward Elgar Publishing. https://EconPapers.repec.org/RePEc:elg:eechap:20820_10

Bennett, P. (2021). *Marketing effectiveness: How to measure it & present to external stakeholders*. HubSpot. Retrieved 24 August from https://blog.hubspot.com/marketing/easy-ways-to-measure-the-effectiveness-of-your-content

Boldt, L. C., Vinayagamoorthy, V., Winder, F., Schnittger, M., Ekran, M., Mukkamala, R. R., Lassen, N. B., Flesch, B., Hussain, A. & Vatrapu, R. (2016). "Forecasting Nike's sales using Facebook data", *2016 IEEE International Conference on Big Data (Big Data)*, pp. 2447–2456.

Bouchet, A., Troilo, M., & Walkup, B. R. (2016). Dynamic pricing usage in sports for revenue management. *Managerial Finance, 42*(9), 913–921. https://doi.org/10.1108/MF-01-2016-0017

Boulton, C. (2018). *Golden State Warriors shoot to augment fan experience with digital tech*. IDG Communications, Inc. Retrieved 25 August from https://www.cio.com/article/228899/golden-state-warriors-shoot-to-augment-fan-experience-with-digital-tech.html

Carrillat, F. A., & d'Astous, A. (2012). The sponsorship-advertising Interface: Is less better for sponsors? *European Journal of Marketing*. https://doi.org/10.1108/03090561211202611

Chaffey, D., & Smith, P. R. (2022). *Digital marketing excellence: Planning, Optimizing and Integrating Online Marketing*. Taylor & Francis. https://books.google.co.uk/books?id=GmmdEAAAQBAJ

Chmait, N., & Westerbeek, H. (2021). Artificial intelligence and machine learning in sport research: An introduction for non-data scientists [perspective]. *Frontiers in Sports and Active* Living, 3. https://www.frontiersin.org/articles/10.3389/fspor.2021.682287

Chmait, N., Westerbeek, H., Eime, R., Robertson, S., Sellitto, C., & Reid, M. (2020). Tennis influencers: The player effect on social media engagement and demand for tournament attendance. *Telematics and Informatics, 50*, 101381. https://doi.org/10.1016/j.tele.2020.101381

Courty, P., & Davey, L. (2019). The impact of variable pricing, dynamic pricing, and sponsored secondary markets in major league baseball. *Journal of Sports Economics, 21*(2), 115–138. https://doi.org/10.1177/1527002519867367

Davenport, T. H., & Harris, J. G. (2007). *Competing on analytics: The new science of winning*. Harvard Business School Press. https://books.google.co.uk/books?id=n7Gp7Q84hcsC

de Vries, N. J., Carlson, J., & Moscato, P. (2019). Clustering consumers and cluster-specific Behavioural models. In P. Moscato & N. J. de Vries (Eds.), *Business and consumer analytics: New ideas* (pp. 235–267). Springer International Publishing. https://doi.org/10.1007/978-3-030-06222-4_5

Dees, W., Walsh, P., McEvoy, C. D., McKelvey, S., Mullin, B. J., Hardy, S., & Sutton, W. A. (2021). *Sport marketing*. Human Kinetics. https://books.google.com/books?id=bxcjEAAAQBAJ

Di Virgilio, F., Soliman, M., Anwar ul Haq, M, & Fantini, S. (2022). Analysing users' engagement with eSports team: Does Covid-19 matter? In S. R. Mondal, F. Di Virgilio, & S. Das (Eds.), *HR analytics and digital HR practices: Digitalization post COVID-19* (pp. 243–271). Springer Nature Singapore. https://doi.org/10.1007/978-981-16-7099-2_10

Digital Strategy Consulting. (2020). *Case study: Adidas uses Google data insights to create targeted storytelling and more relevant product messages*. Digital Strategy Consulting Ltd. Retrieved 28 August from https://www.digitalstrategyconsulting.com/online-advertising/ad-tech/case-study-adidas-uses-google-data-insights-to-create-targeted-storytelling-and-more-relevant-product-messages/48603/

Doyle, C. (2016). *A dictionary of marketing*. OUP Oxford. https://books.google.co.uk/books?id=lGcRDAAAQBAJ

Eagleman, A. N. (2013). Acceptance, motivations, and usage of social media as a marketing communications tool amongst employees of sport national governing bodies. *Sport Management Review, 16*(4), 488–497. https://doi.org/10.1016/j.smr.2013.03.004

FC Barcelona. (2020). *Barça's digital strategy creates new relationship with fans in order to adapt to changing consumer habits*. FC Barcelona. Retrieved 21 August from https://www.fcbarcelona.com/en/news/1601557/barcas-digital-strategy-creates-new-relationship-with-fans-in-order-to-adapt-to-changing-consumer-habits

Fenton, A., Keegan, B. J., & Parry, K. D. (2021). Understanding sporting social media brand communities, place and social capital: A netnography of football fans. *Communication & Sport, 11*(2), 313–333. https://doi.org/10.1177/2167479520986149

Fullerton, S. (2021). *Sports marketing*. SAGE Publications. https://books.google.co.uk/books?id=tsq8EAAAQBAJ.

Giorgio, P. (2017). *How data can help drive sports sponsorship and fan engagement*. Deloitte. Retrieved 24 August from https://www2.deloitte.com/content/dam/Deloitte/us/Documents/consumer-business/us-cb-sports-data-sponsorship-fan-engagement.pdf

Giulianotti, R. (2015). *Sport: A Critical Sociology*, Polity Press, Cambridge.

Giusti Gestri, L. (2023). Wearable technology may assist in reducing jockeys' injuries if integrated into their safety vests: A qualitative study [Original research]. *Frontiers in Sports and Active Living* 5. https://www.frontiersin.org/articles/10.3389/fspor.2023.1167110

Glebova, E., Gerke, A., & Book, R. (2023). The transformational role of technology in sports events. In B. Basu, M. Desbordes, & S. Sarkar (Eds.), *Sports management in an uncertain environment* (pp. 169–187). Springer Nature Singapore. https://doi.org/10.1007/978-981-19-7010-8_8

Goff Inglis, K., & Zolfaghari, S. (2017). A review of scheduling problems and research opportunities in motion picture exhibition. *Interfaces, 47*(2), 175–187. https://doi.org/10.1287/inte.2016.0864

Grant, R. M. (2021). *Contemporary strategy analysis*. Wiley. https://books.google.com/books?id=TadJEAAAQBAJ

Green, F. (2018). *Winning with data: CRM and analytics for the business of sports*. Taylor & Francis. https://books.google.com/books?id=VkpnDwAAQBAJ

Harrison, C. K., & Bukstein, S. (2022). *Sport business analytics: Using data to increase revenue and improve operational efficiency*. Auerbach Publishers, Incorporated. https://books.google.com/books?id=n7EZzwEACAAJ

Hutchins, B., Li, B., & Rowe, D. (2019). Over-the-top sport: Live streaming services, changing coverage rights markets and the growth of media sport portals. *Media, Culture & Society, 41*(7), 975–994. https://doi.org/10.1177/0163443719857623

IBM. (2022). *Putting the FAN in fantasy: How IBM is using artificial intelligence to enhance the ESPN fantasy football experience*. IBM. Retrieved 23 August from https://newsroom.ibm.com/Putting-the-FAN-in-Fantasy-How-IBM-is-Using-Artificial-Intelligence-to-Enhance-the-ESPN-Fantasy-Football-Experience

Jarvis, D., & Westcott, K. (2021). *The hyperquantified athlete: Technology, measurement, and the business of sports*. Deloitte. Retrieved 24 August from https://www2.deloitte.com/xe/en/insights/industry/technology/technology-media-and-telecom-predictions/2021/athlete-data-analytics.html

Jensen, J. A. (2023). Searching for the "Holy Grail" of sponsorship-linked marketing: A generalizable sponsorship ROI model. *arXiv preprint arXiv:2305.09473*.

Jin, C.-H. (2017). Retrospection and state of sports marketing and sponsorship research in from 1999 to 2015. *International Journal of Sports Marketing and Sponsorship*. https://doi.org/10.1108/ijsms-04-2016-0002

Johnson, L. (2017). *Golden State Warriors enlists mobile for fan engagement*. Industry Dive. Retrieved 25 August from https://www.marketingdive.com/ex/mobilemarketer/cms/news/content/14644.html

Kennedy, H., Kunkel, T., & Funk, D. C. (2021). Using predictive analytics to measure effectiveness of social media engagement: A digital measurement perspective [report]. *Sport Marketing Quarterly, 30*, 265+. https://link.gale.com/apps/doc/A689860746/AONE?u=anon~839bc2c4&sid=googleScholar&xid=30847ad5

Kern, T. (2020). *How AI driven ticket pricing is increasing attendance and revenue*. MarketScale. Retrieved 27 August from https://marketscale.com/industries/sports-and-entertainment/how-ai-driven-ticket-pricing-is-increasing-attendance-and-revenue/

Kim, D.-H., Ko, Y., Lee, J. L., & Kim, Y. C. (2019). The impact of CSR-linked sport sponsorship on consumers' reactions to service failures. *International Journal of Sports Marketing and Sponsorship*. https://doi.org/10.1108/ijsms-01-2019-0011

Kings Rocket Department. (2021). *Kings partner with LAVA and flow to release first-ever smart ticket NFT experience pass built on Blockchain*. NBA Media Ventures, LLC. Retrieved 29 August from https://www.nba.com/kings/news/kings-partner-lava-and-flow-release-first-ever-smart-ticket-nft-experience-pass-built

Kozinets, R. V. (2002). The field behind the screen: Using netnography for marketing research in online communities. *Journal of Marketing Research, 39*(1), 61–72. https://doi.org/10.1509/jmkr.39.1.61.18935

Kozinets, R. V. (2019). *Netnography: The Essential Guide to Qualitative Social Media Research*, SAGE Publications, London.

Kramer, K., Wagner, D., & Scheck, B. (2021). Reaping the digital dividend? Sport marketing's move into esports: Insights from Germany. *European Journal of International Management, 15*(2–3), 339–366. https://doi.org/10.1504/EJIM.2021.113265

Krüger, Z. (2023). Creating compelling content. In Z. Krüger (Ed.), *The art of SXO : Placing UX design methods into SEO best practices* (pp. 127–194). Apress. https://doi.org/10.1007/978-1-4842-9212-9_3

LaLoggia, J. (2019). *How the Yankees created a dynamic analytics department: The Yankees went from payroll to analytics juggernaut over the past decade*. Vox Media, LLC. Retrieved 27 August from https://www.beyondtheboxscore.com/2019/4/4/18294438/mlb-yankees-data-analytics-brian-cashman-steinbrenner-chapman-betances-voit-ottavino-world-series

Lim, S. (2023). *How OTT networks are revolutionising sports media and brand sponsorships*. Haymarket Media Ltd.. Retrieved 28 August from https://www.campaignasia.com/article/how-ott-networks-are-revolutionising-sports-media-and-brand-sponsorships/483328

Liu, H., Tan, K. H., & Pawar, K. (2022). Predicting viewer gifting behavior in sports live streaming platforms: The impact of viewer perception and satisfaction. *Journal of Business Research, 144*, 599–613. https://doi.org/10.1016/j.jbusres.2022.02.045

López-Carril, S., Escamilla-Fajardo, P., González-Serrano, M. H., Ratten, V., & González-García, R. J. (2020). The rise of social media in sport: A bibliometric analysis. *International Journal of Innovation and Technology Management, 17*(06), 2050041. https://doi.org/10.1142/S0219877020500418

Lough, N., & Geurin, A. N. (2019). Routledge handbook of the business of women's sport. *Taylor & Francis*. https://books.google.com/books?id=hbyZDwAAQBAJ

Lyons, J. (2023). *5 essential sports marketing strategies*. Constant Contact Retrieved 25 August from https://www.constantcontact.com/blog/sports-marketing-strategies/

Mahmoud, A. B., Hack-Polay, D., Grigoriou, N., Mohr, I., & Fuxman, L. (2021). A generational investigation and sentiment and emotion analyses of female fashion brand users on Instagram in Sub-Saharan Africa. *Journal of Brand Management, 28*(5), 526–544. https://doi.org/10.1057/s41262-021-00244-8

Mahmoud, A. B., Fuxman, L., Asaad, Y., & Solakis, K. (2024). Exploring new realms or losing touch? Assessing public beliefs about tourism in the metaverse–a big-data approach. *International Journal of Contemporary Hospitality Management*. https://doi.org/10.1108/IJCHM-09-2023-1515

Mahmoud, A. B. (2024a) Aging in the digital age: Public beliefs about the potential of virtual reality (VR) for the aging population. *Journal of Macromarketing*. https://doi.org/10.1177/02761467241253286

Mahmoud, A. B. (2024b). Exploring the public's beliefs, emotions and sentiments towards the adoption of the metaverse in education: A qualitative inquiry using big data. *British Educational Research Journal*. https://doi.org/10.1002/berj.4026

Mamo, Y., Su, Y., & Andrew, D. P. S. (2022). The transformative impact of big data applications in sport marketing: Current and future directions. *International Journal of Sports Marketing and Sponsorship, 23*(3), 594–611. https://doi.org/10.1108/IJSMS-03-2021-0073

Man Utd Communications Department. (2022). *Man Utd appoints head of fan engagement*. Manchester United FC Ltd. Retrieved 21 August from https://www.manutd.com/en/news/detail/man-utd-appoints-head-of-fan-engagement-15-february-2022

Maricic, M., Kostic-Stankovic, M., Bulajic, M., & Jeremic, V. (2019). See it and believe it? Conceptual model for exploring the recall and recognition of embedded advertisements of sponsors. *International Journal of Sports Marketing and Sponsorship, 20*(2), 333–352. https://doi.org/10.1108/IJSMS-02-2018-0020

Mathew, L., & George, S. (2022). The Indian fitness industry: Impact of digitalisation. In J. García-Fernández, M. Valcarce-Torrente, S. Mohammadi, & P. Gálvez-Ruiz (Eds.), *The digital transformation of the fitness sector: A global perspective* (pp. 97–109). Emerald Publishing Limited. https://doi.org/10.1108/978-1-80117-860-020221014

McCarthy, J., Rowley, J., & Keegan, B. J. (2022). Social media marketing strategy in English football clubs. *Soccer & Society, 23*(4–5), 513–528. https://doi.org/10.1080/14660970.2022.2059872

McLaughlin, M. (2018). *Sports data analytics: How teams use data to gain edge: Teams and athletes gain a better understanding of how to win with insights unearthed by technology*. CDW LLC. Retrieved 28 August from https://biztechmagazine.com/article/2018/12/how-data-analytics-revolutionizing-sports

Miller, T. W. (2015). *Sports analytics and data science: Winning the game with methods and models*. Pearson Education. https://books.google.com/books?id=7LXxCgAAQBAJ

Mohammadi, S., Heidari, A., & Navkhsi, J. (2023). Proposing a framework for the digital transformation maturity of electronic sports businesses in developing countries. *Sustainability, 15*(16).

Moran, E. (2020). *NBA looks to superserve fans with 'virtual content factory'*. Front Office Sports. Retrieved 24 August from https://frontofficesports.com/nba-virtual-content-factory/

Mullin, R. (2018). *Promotional marketing*. Taylor & Francis. https://books.google.co.uk/books?id=Rb9aDwAAQBAJ

Mumcu, C. (2019). Business analytics in women's professional sports. In *Routledge handbook of the business of women's sport* (pp. 239–251). https://doi.org/10.4324/9780203702635-19

Murray, P. (2021). *Inside the golden state warriors' virtual fan experience*. Forbes. Retrieved 25 August from https://www.forbes.com/sites/patrickmurray/2021/01/10/inside-the-golden-state-warriors-virtual-fan-experience/

Murthy, D. (2011). Twitter: Microphone for the masses? *Media, Culture & Society, 33*(5), 779–789. https://doi.org/10.1177/0163443711404744

Naraine, M. L., Wear, H. T., & Whitburn, D. J. (2019). User engagement from within the Twitter community of professional sport organizations. *Managing Sport and Leisure, 24*(5), 275–293. https://doi.org/10.1080/23750472.2019.1630665

Naraine, M. L., Hayduk, T., & Doyle, J. P. (2022). *The Routledge handbook of digital sport management*. Taylor & Francis. https://books.google.co.uk/books?id=gICUEAAAQBAJ

NBA. (2020). *Golden State Warriors virtual stadium tour*. NBA Media Ventures, LLC. Retrieved 29 August from http://warriors.io-media.com/web/index.html

O'Hallarn, B., Shapiro, S. L., Wittkower, D. E., Ridinger, L., & Hambrick, M. E. (2019). A model for the generation of public sphere-like activity in sport-themed Twitter hashtags. *Sport Management Review, 22*(3), 407–418. https://doi.org/10.1016/j.smr.2018.06.001

olympics.com. (2019). *The IOC, the Coca-Cola company and China Mengniu dairy company ltd announce joint worldwide Olympic partnership to 2032*. International Olympic Committee. Retrieved 27 August from https://olympics.com/ioc/news/the-ioc-the-coca-cola-company-and-china-mengniu-dairy-company-ltd-announce-joint-worldwide-olympic-partnership-to-2032

Oyewole, G. J., & Thopil, G. A. (2023). Data clustering: Application and trends. *Artificial Intelligence Review, 56*(7), 6439–6475. https://doi.org/10.1007/s10462-022-10325-y

Padmanabhan, B., & Tuzhilin, A. (2003). On the use of optimization for data mining: Theoretical interactions and eCRM opportunities. *Management Science, 49*(10), 1327–1343. https://doi.org/10.1287/mnsc.49.10.1327.17310

Pappalardo, L., Rossi, A., Natilli, M., & Cintia, P. (2021). Explaining the difference between men's and women's football. *PLoS One, 16*(8), e0255407. https://doi.org/10.1371/journal.pone.0255407

Pattanayak, S., Pati, P. B., & Singh, T. (2022, 27–29 May 2022). Performance analysis of machine learning algorithms on multi-touch attribution model. *2022 3rd International Conference for Emerging Technology (INCET)*

Peymandoust, A. (2021). *How AI is helping Adidas boost its personalized email marketing game*. Salesforce, Inc. Retrieved 28 August from https://www.salesforce.com/blog/ai-boosts-adidas-personalized-email-marketing/

Pradeep, A. K., Appel, A., & Sthanunathan, S. (2018). *AI for marketing and product innovation: Powerful new tools for predicting trends, connecting with customers, and closing sales*. Wiley. https://books.google.co.uk/books?id=pC5zDwAAQBAJ

Pykes, K. (2022). *Sports analytics: How different sports use data analytics*. DataCamp. Retrieved 28 August from https://www.datacamp.com/blog/sports-analytics-how-different-sports-use-data-analysis

Quansah, T. K., Buraimo, B., & Lang, M. (2023). Determining the price of football: An analysis of matchday ticket prices in the English premier league. *European Sport Management Quarterly, 1-21*. https://doi.org/10.1080/16184742.2023.2191633

Rascher, D. A., McEvoy, C. D., Nagel, M. S., & Brown, M. T. (2007). Variable ticket pricing in major league baseball. *Journal of Sport Management, 21*(3), 407–437.

Read, D., & Smith, A. C. T. (2023). Data as capital and ethical implications in digital sport business models. *Convergence, 13548565231176384*. https://doi.org/10.1177/13548565231176384

Rogic, S., & Kascelan, L. (2019). *Customer value prediction in direct marketing using hybrid support vector machine rule extraction method*. New Trends in Databases and Information Systems.

Schroer, A. (2022). *How sports analytics are used today, by teams and fans*. Built In. Retrieved 28 August from https://builtin.com/big-data/big-data-companies-sports

Schwarz, E. C., & Hunter, J. D. (2017). *Advanced theory and practice in sport marketing*. Routledge.

Schwarz, E. C., Brannigan, K. J., Hunter, J. D., & Cattani, K. P. (2022). *Advanced theory and practice in sport marketing*. Routledge. https://books.google.co.uk/books?id=DWvmzgEACAAJ

Scott, G. (2018). *5 ways sports teams use CRM systems*. BIZTECH. Retrieved 27 August from https://biztechmagazine.com/article/2018/11/5-ways-sports-teams-use-crm-systems

Šerić, N., & Ljubica, J. (2018). *Market research methods in the sports industry*. Emerald Publishing Limited. https://doi.org/10.1108/9781787541917

Shabbirhusain, R. V., Annamalai, B., & Chandrasekaran, S. (2023). Global multi-sport events: Content strategy for driving fan engagement on Twitter. *Sport, Business and Management: An International Journal, 13*(4), 450–469. https://doi.org/10.1108/SBM-10-2022-0094

Sim, J. (2023). *NBA delivers record digital and social engagement for 2022/23 regular season*. SportsPro. Retrieved 25 August from https://www.sportspromedia.com/news/nba-digital-social-fan-engagement-instagram-youtube-league-pass-app-2022-23/?zeplir_sso_ott=n9JdV1

Singh, D. S. P. (2023). Role and impact of E-banking on digital marketing. *Journal of Economy, Tourism and Service, 2*(7), 39–59. http://jets.innovascience.uz/index.php/jets/article/view/69

Smith, A. C. T., & Stewart, B. (2010). The special features of sport: A critical revisit. *Sport Management Review, 13*(1), 1–13. https://doi.org/10.1016/j.smr.2009.07.002

Socios. (2022). *FC Barcelona to accelerate Web3 strategy as Socios.com invests $100m in Barca studios*. Mediarex Enterprises Limited. Retrieved 21 August from https://www.socios.com/fc-barcelona-to-accelerate-web3-strategy-as-socios-com-invests-100m-in-barca-studios/

Srivastav, T. (2018). *Vivoom, Viber and Dugout join forces to drive FC Barcelona's digital fan engagement and brand safety*. Carnyx Group Ltd.. Retrieved 21 August from https://www.thedrum.com/news/2018/10/31/vivoom-viber-and-dugout-join-forces-drive-fc-barcelonas-digital-fan-engagement-and

StriveCloud. (2022). *Engagement breakdown: How Manchester United is doubling down on digital fan engagement*. StriveCloud. Retrieved 21 August from https://strivecloud.io/blog/fan-engagement/manu-digital-engagement-lockdown/

The Stadium Business. (2020). *Barcelona progresses AI-led 'digital twin' crowd management project*. TheStadiumBusiness. Retrieved 29 August from https://www.thestadiumbusiness.com/2020/11/13/barcelona-progresses-ai-led-digital-twin-crowd-management-project/

Tucker, T. (2016). LEADOFF: Falcons sell $900M in stadium sponsorships, report says. *The Atlanta Journal-Constitution*. Retrieved 28 August from https://www.ajc.com/sports/football/leadoff-falcons-sell-900m-stadium-sponsorships-report-says/KCi2AKCazjeWojJzxNqXCJ/

Valendin, J., Reutterer, T., Platzer, M., & Kalcher, K. (2022). Customer base analysis with recurrent neural networks. *International Journal of Research in Marketing, 39*(4), 988–1018. https://doi.org/10.1016/j.ijresmar.2022.02.007

van Dam, J.-W., & van de Velden, M. (2015). Online profiling and clustering of Facebook users. *Decision Support Systems, 70*, 60–72. https://doi.org/10.1016/j.dss.2014.12.001

Venkatesan, R., Farris, P. W., & Wilcox, R. T. (2021). *Marketing analytics: Essential tools for data-driven decisions*. University of Virginia Press. https://books.google.com/books?id=3KsBEAAAQBAJ

Vitale, M. (2021). *Global sports sponsorships are now purpose-driven marketing*. Forbes. Retrieved 28 August from https://www.forbes.com/sites/forbesagencycouncil/2021/06/11/global-sports-sponsorships-are-now-purpose-driven-marketing/

Westney, A. (2013). *SportVU tracking system could find reach beyond court for NBA, teams*. Leaders Group. Retrieved 21 August from https://www.sportsbusinessjournal.com/Daily/Issues/2013/10/29/NBA-Season-Preview/Tech.aspx

Wittenberg, M., & Meaney, D. (2022). *Stadium analytics: Increasing sports fan engagement with data and AI*. Databricks. Retrieved 27 August from https://www.databricks.com/blog/2022/03/31/stadium-analytics-increasing-sports-fan-engagement-with-data-and-ai.html

Wood, J., & Smith, R. (2018). *Mercedes-Benz stadium*. Dimensional Innovations. Retrieved 28 August from https://dimin.com/work/mercedes-benz-stadium

Wood, L., Hoeber, O., Snelgrove, R., & Hoeber, L. (2019). Computer science meets digital leisure: Multiple perspectives on social media and eSport collaborations. *Journal of Leisure Research, 50*(5), 425–437. https://doi.org/10.1080/00222216.2019.1594466

Yadav, J., Misra, M., Rana, N. P., Singh, K., & Goundar, S. (2022). Netizens' behavior towards a blockchain-based esports framework: A TPB and machine learning integrated approach. *International Journal of Sports Marketing and Sponsorship, 23*(4), 665–683. https://doi.org/10.1108/IJSMS-06-2021-0130

Yadav, J., Yadav, R., Sahore, N., & Mendiratta, A. (2023). Digital social engagements and knowledge sharing among sports fans: Role of interaction, identification, and interface. *Technological Forecasting and Social Change, 195*, 122792. https://doi.org/10.1016/j.techfore.2023.122792

Yan, W., Jiang, X., & Liu, P. (2023). A review of basketball shooting analysis based on artificial intelligence. *IEEE Access, 11*, 87344–87365. https://doi.org/10.1109/ACCESS.2023.3304631

Yılmaz, Ö., Easley, R. F., & Ferguson, M. E. (2023). The future of sports ticketing: Technologies, data, and new strategies. *Journal of Revenue and Pricing Management, 22*(3), 219–230. https://doi.org/10.1057/s41272-023-00427-9

Zachlod, C., Samuel, O., Ochsner, A., & Werthmüller, S. (2022). Analytics of social media data—State of characteristics and application. *Journal of Business Research, 144*, 1064–1076. https://doi.org/10.1016/j.jbusres.2022.02.016

Zagnoli, P., & Radicchi, E. (2022). Disruptive technologies on sport event marketing: New centrality of digital communication. In *Marketing analysis in sport business* (pp. 51–69). Routledge.

Zhu, C., Wu, J., Yan, J., & Liu, X. (2023). Advanced fiber materials for wearable electronics. *Advanced Fiber Materials, 5*(1), 12–35. https://doi.org/10.1007/s42765-022-00212-0

Unlocking Athletic Potential: Harnessing People Analytics for Optimal Performance and Talent Management in Sports

Ali B. Mahmoud, William D. Reisel, Alexander Berman, and Leonora Fuxman

1 Introduction

Over the last few years, the sports industry has undergone a significant transformation, largely due to the emergence and widespread adoption of people analytics (Giorgio & Westcott, 2023; Jarvis & Westcott, 2021). This exciting framework, which sits at the intersection of data science and human resource management, may be considered to have the potential to overhaul the way organizations identify talent, enhance player performance, prevent injuries, and coach athletes. At the center of this transformation is the use of vast amounts of sporting data that can provide new insights to team managers and coaches that can push decision-making from intuitive processes to sounder quantitative based decision-making.

The impact of people analytics on sports supersedes player statistics. It represents a comprehensive approach to optimizing human capital within sports organizations, creating a data-driven decision-making culture. The applications of this approach are numerous, from improving recruitment strategies to developing individualized training regimes and even shaping broader organizational values around performance and success.

This chapter contributes to the sports analytics discipline by analyzing examples from the sports industry and discussing emerging trends to provide a comprehensive understanding of how people analytics can be used to unlock athletic potential and promote ethical decision-making within sports organizations. The chapter also

A. B. Mahmoud (✉)
St. John's University, Queens, NY, USA

University of London, London, UK
e-mail: mahmouda@stjohns.edu

W. D. Reisel · A. Berman · L. Fuxman
St. John's University, Queens, NY, USA
e-mail: reiselw@stjohns.edu; bermana@stjohns.edu; fuxmanl@stjohns.edu

A. Mansurali et al. (eds.), *Sports Analytics*,
https://doi.org/10.1007/978-3-031-63573-1_5

examines the ethical implications of using data, highlighting the importance of maintaining a balance between data privacy and performance analysis and addressing potential biases that arise when using data to make decisions.

2 Foundations of People Analytics in Sports

2.1 Defining People Analytics

The trajectory of sports analytics has witnessed a significant evolution from rudimentary statistics to sophisticated analytical models and machine-learning algorithms (Sharma et al., 2023). Initially, sports analytics was primarily concerned with basic performance metrics and historical data analysis. However, with technological advancements and the increasing availability of diverse data sources, it has matured into a multi-faceted discipline (Burk & Miner, 2023; Butterworth, 2023; Seçkin et al., 2023). People analytics represents a further evolution, emphasizing a holistic view of the athlete, encompassing not just performance but a myriad of factors that contribute to athletic success and organizational effectiveness in the sports domain (Araújo et al., 2021; Lotfy et al., 2023). People analytics is closely related to HR analytics or talent analytics. It is an empirical approach aimed at making informed decisions regarding talent management, performance optimization, and organizational effectiveness by leveraging data and analytical models (McCartney & Fu, 2022; Suri & Lakhanpal, 2022; Tursunbayeva et al., 2018). In the sports arena, people analytics extends to the systematic collection, analysis, and interpretation of data related to athletes' performance, health, and teamwork dynamics (Chhabra et al., 2023; Fransen et al., 2020; Vella et al., 2022). This form of analytics enables stakeholders to derive actionable insights to enhance both individual and collective outcomes within the sporting ecosystem (Foster et al., 2021).

2.2 Key Metrics and Data Sources

The range of metrics utilized in sports people analytics is vast and encompasses physiological, biomechanical, psychological, and tactical dimensions. Key metrics include players' physical attributes (e.g., speed, strength, agility), performance statistics (e.g., goals scored, assists, defensive actions), and health metrics (e.g., injury history, recovery rates). Data sources range from traditional sources like game statistics to more advanced tools like wearables, tracking systems, and even genomic data, which are becoming increasingly prevalent. For instance, the Minnesota Twins Baseball Team worked with Databricks to run up to 20,000 simulations for each historical pitch they have on a player in near real-time. This helps them better understand the likelihood of success for particular batters based on the pitcher, pitch rotation, speed, placement, and weather conditions (Leano, 2021). Another example is Real Madrid Football Club which uses sports analytics for predicting and preventing injuries, evaluating for optimal performance, maximizing training utilizing data, and tracking fan behavior (Ahmed, 2023). Furthermore, Billy Beane, the Oakland Athletics General Manager popularized in Moneyball, used market inefficiencies to his advantage. He prioritized sabermetrics such as on-base percentage to identify players he could sign at low salaries and turn his team into contenders for the World Series (Eric, 2023).

3 Talent Identification and Recruitment

3.1 Assessing and Predicting Athlete Potential

The importance of sports talent identification and development has grown over the last two decades (Leite et al., 2021; Sarmento et al., 2018). The focal in this analysis is understanding the inherent qualities that propel elite performance in athletes and the part coaches play in nurturing enhanced performance (Leite et al., 2021). Factors such as relative age, growth, maturation, and training age are believed to be crucial in talent identification models (Kelly, 2023; Salter et al., 2023). Athlete evaluation models place a higher value on current performance, which can underestimate the athlete's physical and maturation progress that predicts future performance (Thornton et al., 2019; Till & Baker, 2020; Vaeyens et al., 2008). Current research is shifting focus to a more holistic appraisal of talent, encompassing anthropometric, physiological, psychological, and contextual elements, striving to ameliorate the overarching process of talent identification and development (Thornton et al., 2019). In this regard, neuro assessment in sports and athlete analytics involves the use of neurodiagnostic tools to evaluate an athlete's brain processing during sports-related movements. This includes non-invasive brain imaging techniques like functional near-infrared spectroscopy (fNIRS) and targeted non-invasive brain stimulation such as transcranial direct current stimulation (tDCS), where these techniques can help optimize training outcomes and guide training processes (Seidel-Marzi & Ragert, 2020). Additionally, computerized cognitive assessment tests are used for detecting mild cognitive dysfunction in athletes with sports-related head injuries (Collie et al., 2001). The utilization of neuro assessment in sports provides an integrative approach for performance enhancement, where data helps profile athletes' strengths and weaknesses, monitor progress, adapt training protocols, and, importantly, identify or predict an athlete's potential (Crivelli & Balconi, 2022).

3.2 Data-Driven Scouting Techniques

Data analysis has notably transformed scouting in football. Football clubs, as well as independent and freelance scouts, harness the power of data analysis to improve their scouting processes, aiding them in making more informed decisions based on the vast resources available, including data (Pavitt et al., 2021). A specific example showcases how the performance of Ousmane Dembélé, a right winger for Barcelona, is analyzed in different playing situations through data-driven scouting. This kind of analysis helps in understanding a player's capabilities and performance under varied circumstances (Andrianova & Fedoseev, 2021).

3.3 Impact of People Analytics on Recruitment Strategies

Sports analytics relies on data for soccer teams in recent years, aiding in the identification and recruitment of promising players while also enabling the analysis of a team's performance to prevent injuries (Thomas, 2023). The methodologies employed by coaches for in-game decisions and player recruitment have undergone a substantial shift, driven by the insights gleaned from data analytics (Fitzpatrick, 2020). Utilizing current and past data to predict players' performance has gained traction, particularly in basketball. Sports analytics facilitates the evaluation of players and teams from multiple perspectives, thereby informing the recruitment strategies deployed by technical staff and coaches (Sarlis & Tjortjis, 2020). Having a well-defined and thorough understanding of the roles is fundamental for hiring the right individual in sports management. Analyzing the position and defining the responsibilities aids in assessing the work environment the candidates will need to operate in, aligning the recruitment process with the strategic objectives of the sports organization (Hanson, 2023).

4 Player Performance Analysis

Analyzing player performance is a dynamic and multifaceted endeavor, encompassing everything from scrutinizing individual abilities to assessing team interplay. By personalizing coaching and development initiatives, this meticulous analysis sheds light on the inner workings of a team as well as the contributions made by each player. The insights gained through this process are invaluable—they highlight areas of strength and pinpoint potential weaknesses, both for individual players and the team as a whole. These findings can then inform tailored training strategies designed to elevate performance. Furthermore, it's not just about reflecting on what's already happened; predictive analysis uses historical data to forecast future achievements while also gauging player responses to varying scenarios and evaluating injury risks. In essence, understanding these diverse factors is pivotal to refining skills, boosting team synergy, and ultimately steering a sports team toward the pinnacle of success.

Quantitative metrics such as goals scored, assists, passes caught, and strikeouts made have long been used for ranking individual players (Foster et al., 2021). Athlete performance can be quantified and assessed using various methods, some of which include vertical jump tests to determine sport preparedness and actual performance statistics such as the number of kills, blocks, or assists in a match for a volleyball player (Bunn et al., 2020). The collected data from assessments provides direct feedback on an athlete's development and potentially leads to alterations in their training programs. Additionally, from a talent scouting perspective, focusing on athletes who exhibit optimal performance in specific variables is crucial (Bailey, 2021). For instance, data mining algorithms and techniques are employed to evaluate basketball players' performance and to analyze strategies for team composition improvement (Sarlis & Tjortjis, 2020). Furthermore, talent identification programs utilize the assessment of sport-specific technical skills to distinguish between more and less skilled individuals, aiding in predicting future performance (Koopmann et al., 2020).

Advanced statistics like those used in basketball for NBA and Euroleague player and team evaluations have become common across major professional sports globally (Sarlis & Tjortjis, 2020). These analytics provide crucial insights for scouting opponents, identifying team areas for improvement, influencing in-game strategies, guiding training processes, and boosting performance levels. Examples include on-field data analytics in soccer that track metrics to refine methodologies, analyses that helped a football manager decide which goalkeeper to substitute based on penalty-saving percentages, and tennis players leveraging data to adapt positioning against an opponent's favored shots (Pykes, 2022). Beyond sports practitioners, elite teams are also utilizing AI applications to process volumes of data and generate insights that revolutionize data-driven decision-making (Bhatnagar & Babbar, 2022; Houtmeyers et al., 2021; Liu et al., 2023). Whether for substitutions, training regimens, or game-time strategies, data analytics support personalized development plans and provide professional sports teams with a competitive edge.

5 Injury Prevention and Rehabilitation

In the realm of sports, identifying risk factors for injuries through data analysis stands as an indispensable facet of athlete health management (Andrew et al., 2019; West et al., 2020). The application of sports analytics extends not only to assessing an athlete's prospective performance but also to their susceptibility to injuries. For instance, orthopedic surgeons at Mayo Clinic collaborate with professional athletes to identify ways to circumvent potential risks and devise optimal treatment plans for injuries when they occur (Mayo Clinic, 2021). Additionally, machine learning methods are deployed to detect athletes at high risk of injuries and find risk factors, offering a proactive injury prevention plan (Van Eetvelde et al., 2021). Wearable technologies, when aggregated with analytics, further augment this endeavor by identifying injury risk factors (Zadeh et al., 2021), thereby channeling focus toward risk reduction and contributing to a safer sporting environment.

Data-driven strategies to curb injuries serve as a prudent approach to safeguard athlete well-being. Machine learning methods are harnessed to enhance injury prediction accuracy, thus facilitating timely preventative measures (Van Eetvelde et al., 2021). Sports scientists advocate for the deployment of multiple preventative methodologies in tandem—such as preventative exercises, training load management, squad rotation, and recovery activities—to significantly diminish the likelihood of musculoskeletal injuries (Tee et al., 2020). The efficacy of such preventative programs is substantiated by many comprehensive reviews and published meta-analyses (e.g., Stephenson et al., 2021), underscoring the pivotal role of analytics in injury prevention strategies in sports.

Analytics can also assist in recovery following injuries, leveraging targeted rehabilitation and recovery strategies (Owens et al., 2020). Mobile rehabilitation (mRehab), through sensor-enhanced activity management (SEAM), has enabled the creation of personalized home exercise regimens, with therapists fine-tuning exercise progressions based on clinic assessments (Jones et al., 2020). Moreover, the burgeoning field of Rehabilomics, supported by machine learning, aims at functional recovery prediction to enhance the scientific substantiation of rehabilitation protocols and tailor medical therapies (Campagnini et al., 2022). This represents a significant stride toward harnessing analytics to foster a data-driven, personalized approach to athlete rehabilitation and recovery, thereby underscoring the monumental potential of analytics in sports medicine.

6 Coaching and Leadership Development

In the sphere of sports, coaching effectiveness significantly influences athlete performance and overall team success (Smith & Smoll, 2017). Assessing and enhancing coaching effectiveness have thus become critical, with various aspects being analyzed to ascertain impactful coaching practices. For instance, the influence of coaching interventions on outcomes like team cohesion, self-esteem, and performance anxiety is meticulously studied to understand the broader impact of coaching strategies (Foulds et al., 2019). Additionally, the perception of the coach–athlete relationship is explored to gauge its bearing on coaching effectiveness, recognizing that a positive rapport can promote athlete satisfaction and improved performance (Mohd Kassim et al., 2020). Furthermore, evaluating the effectiveness of coaching techniques is deemed essential to foster enhanced communication between athletes and coaches, which, in turn, can lead to refined performance and heightened satisfaction among athletes (Davis et al., 2019; Jowett, 2017; Kao & Tsai, 2016).

Leadership qualities in athletes play a crucial role in guiding teams toward success (Cotterill et al., 2022). The distinct leadership attributes of athletes, encompassing accountability, patience, integrity, and vision, are recognized as pivotal elements that cultivate sportsmanship, trust, credibility, and team unity (Santosh, 2021). Formal and informal athlete leaders, including team captains, significantly contribute to the team's success by embodying these qualities and developing team cohesion (Buchanan & Schlegel, 2019).

Building cohesive teams is a multi-faceted process that greatly benefits from the insights derived from people analytics. The success stories of smaller countries like Iceland and Wales in the 2016 Football European Championships underscore the importance of team cohesion, which helped them overcome deficiencies in individual talent (Eys & Kim, 2017). Effective leadership is identified as a strong correlate to team cohesiveness, where leaders' approaches significantly impact teamwork, innovation, and overall team performance (Grossman et al., 2021). For instance, team coordination training, such as Crew Resource Management (CRM), is designed to improve teamwork by educating team members on effective team coordination and communication, thereby contributing to enhanced team effectiveness and cohesiveness (Shuffler et al., 2018). Through the lens of people analytics, the exploration of these facets—coaching effectiveness, athlete leadership qualities, and team cohesion—provides a holistic approach to nurturing a conducive environment for superior performance and success in the sporting arena.

7 Ethical Considerations and Challenges

The application of people analytics in the sports sector raises crucial ethical considerations related to data privacy concerns when seeking performance analysis. As performance analysis grows rapidly, tackling issues such as data rights management and privacy while concurrently harnessing analytics for performance augmentation is imperative. The focal point is thoughtfully managing data rights and privacy, ensuring an ethical foundation for evaluating performance (Jarvis & Westcott, 2021).

The second critical ethical consideration is the potential biases inherent in sports performance evaluations. The subjective tendencies of evaluators could inadvertently introduce biases, thus skewing judgments and analyses. For instance, overestimations of performances by competitors from one's own country or the overvaluation of certain stylistic elements could result in biased assessments (Lyngstad et al., 2020; Plessner & Haar, 2006). This suggests the need for neutral, unbiased evaluation systems, especially when analytics are employed to inform decisions pertaining to athlete performance and recruitment (Lyngstad et al., 2020).

Lastly, the principle of fairness and transparency in decision-making is highlighted as AI technologies revolutionize people analytics. Ethical challenges such as biased data and lack of transparency in AI algorithms necessitate robust ethical frameworks (Najjar, 2023; Orlando, 2022). Having ethical committees, regular bias testing in algorithms, and adherence checks with people analytics technology vendors are keys. Such deterrents foster fairness, transparency, and unbiased decision-making, exemplifying the ethical principles necessary for the responsible use of people analytics in sports, thereby ensuring a balanced approach to innovation and the ethical utilization of AI technologies (Randhawa, 2023).

8 The Future of People Analytics in Sports

A notable trend is the substantial growth of the sports analytics market, expected to reach $8.4 billion by 2026 from $2.5 billion in 2021, reflecting the escalating interest and investments in this field (Friend, 2022). This growth is fueled by advancements in cloud technology, which facilitates seamless data storage and real-time analytics. Sports entities like Formula One and Real Madrid are leveraging cloud solutions to meet their data analytics requirements (Friend, 2022). Moreover, the advent of Artificial Intelligence (AI) and Machine Learning (ML) has further monetized the analytics landscape. These technologies are pivotal in extracting actionable insights from data, including player performance predictions and injury prevention, backed by a variety of learning algorithms like statistical learning and deep learning (Fury et al., 2022; Ghosh et al., 2023; Khanna, 2023).

The application of these emerging technologies fosters numerous benefits, especially in the realms of player development, in-game strategy, and health and performance monitoring. For instance, analytics play a crucial role in driving player development and strategizing during games, where rapid data access can significantly alter game outcomes (Al-Asadi & Tasdemır, 2022; Chowdhary et al., 2023; Friend, 2022; Huyghe et al., 2022). Additionally, continuous monitoring of athletes' health and performance through technologies transforms how athletes train, compete, and manage their careers (Pickering & Kiely, 2019; Strength et al., 2021). This data-driven approach extends to enhancing coaching and training regimes, where analytics aid in design, monitoring, and evaluation, thereby improving the overall performance of players and teams (Andreassen et al., 2019; Li & Xu, 2021; Patel et al., 2020; Pino-Ortega et al., 2021; Torres-Ronda et al., 2022).

Preparation for the next wave of innovation in athlete analytics underscores the importance of ethical considerations and interdisciplinary collaboration (Li et al., 2023; Otto et al., 2016). The surge in athlete data necessitates a responsible utilization framework to address ethical concerns. Moreover, fostering a collaborative environment between sports medicine researchers and sports science teams is recommended to validate and interpret emerging technologies and big data (Fury et al., 2022; Montull et al., 2022). Such collaborative activities are likely to spur further innovations, ensuring that the sports sector is well-poised to harness the full potential of analytics for enhanced performance and strategic advantages.

9 Conclusion

In summary, this chapter has provided a comprehensive overview of the emerging field of people analytics and its tremendous potential to transform talent management, performance optimization, injury prevention, and coaching in the sports domain. The theoretical foundations were established by defining people analytics as a data-driven approach to inform strategic talent decisions and drive a high-performance culture. Key metrics and data sources were outlined, spanning physical, physiological, psychological, and tactical attributes obtained from advanced tracking technologies and wearables.

The chapter then delved into practical, real-world applications across critical aspects of sports management. For talent identification, the importance of taking a holistic view of potential was emphasized, going beyond current performance: data-driven scouting techniques and the impact of analytics on recruitment strategies exemplified how data is revolutionizing talent spotting. Analytics has enabled a multivariate analysis of individual skills, team dynamics, and personalized development plans. This has helped to evaluate player performance in a comprehensive manner. The role of predictive modeling in injury prevention and the use of data to facilitate rehabilitation further revealed the benefits of an analytical approach. Analytics can play an essential role in coaching, from evaluating interventions to developing leadership skills. By using analytics, coaches can create the ideal conditions for success.

Moreover, the chapter addressed salient ethical considerations, chiefly regarding data privacy and bias mitigation when leveraging analytics for decision-making. It also touched upon emerging technologies and trends catalyzing innovation in this arena, from cloud computing and AI to cross-disciplinary collaborations.

In conclusion, people analytics harbors great potential to transform sports management and performance optimization if deployed judiciously. In this chapter, we were given a comprehensive overview of the numerous applications of sports technology across the sporting ecosystem, along with a thorough examination of the practical and ethical considerations that arise in this field. The sports industry is on the brink of a data revolution, and this chapter offers valuable insights and guidance to leverage analytics in an ethical and effective manner.

References

Ahmed, N. (2023). *Understanding sports data analytics simplified.* Hevo. Retrieved 27 October from https://hevodata.com/learn/sports-data-analytics/

Al-Asadi, M. A., & Tasdemır, S. (2022). Predict the value of football players using FIFA video game data and machine learning techniques. *IEEE Access, 10*, 22631–22645. https://doi.org/10.1109/ACCESS.2022.3154767

Andreassen, K., Johansen, D., Johansen, H., Baptista, I., Pettersen, S. A., Riegler, M., & Halvorsen, P. (2019). Real-time analysis of physical performance parameters in elite soccer. *2019 International Conference on Content-Based Multimedia Indexing (CBMI)* (4–6 Sept. 2019)

Andrew, D. P. S., Pedersen, P. M., & McEvoy, C. D. (2019). *Research methods and design in sport management*. Human Kinetics. https://books.google.com/books?id=9LG8DwAAQBAJ

Andrianova, R. I., & Fedoseev, D. V. (2021). Digital scouting technologies in game sports. In E. G. Popkova, V. N. Ostrovskaya, & A. V. Bogoviz (Eds.), *Socio-economic systems: Paradigms for the future* (pp. 655–662). Springer International Publishing. https://doi.org/10.1007/978-3-030-56433-9_68

Araújo, D., Couceiro, M., Seifert, L., Sarmento, H., & Davids, K. (2021). Artificial intelligence in sport performance. *Analysis*. https://doi.org/10.4324/9781003163589

Bailey, C. A. (2021). *Quantitative analysis in exercise and sport science*. University of North Texas Libraries.

Bhatnagar, R., & Babbar, M. (2022). A systematic review of sports analytics. *International Journal of Technology Transfer and Commercialisation, 19*(4), 393–406. https://doi.org/10.1504/IJTTC.2022.127574

Buchanan, C., & Schlegel, A. (2019). *Leadership in athletics: Transcendence of character in sports*. Ohio State University. Retrieved 27 October from https://fisher.osu.edu/blogs/leadreadtoday/blog/leadership-in-athletics-transcendence-of-character-in-sports

Bunn, J. A., Ryan, G. A., Button, G. R., & Zhang, S. (2020). Evaluation of strength and conditioning measures with game success in division I collegiate volleyball: A retrospective study. *The Journal of Strength & Conditioning Research, 34*(1) https://journals.lww.com/nsca-jscr/fulltext/2020/01000/evaluation_of_strength_and_conditioning_measures.21.aspx

Burk, S., & Miner, G. (2023). *It's all analytics, part III: The applications of AI, analytics, and data science*. Taylor & Francis. https://books.google.com/books?id=BOzNEAAAQBAJ

Butterworth, A. (2023). *Professional practice in sport performance analysis*. Taylor & Francis. https://books.google.com/books?id=9lCtEAAAQBAJ

Campagnini, S., Arienti, C., Patrini, M., Liuzzi, P., Mannini, A., & Carrozza, M. C. (2022). Machine learning methods for functional recovery prediction and prognosis in post-stroke rehabilitation: A systematic review. *Journal of Neuroengineering and Rehabilitation, 19*(1), 54. https://doi.org/10.1186/s12984-022-01032-4

Chhabra, G., Kumar, S., Gupta, S., & Nagpal, P. (2023). Predictive and performance analytics in fitness and sport. In G. Chhabra, S. Kumar, S. Gupta, & P. Nagpal (Eds.), *Artificial intelligence to analyze psychophysical and human lifestyle* (pp. 129–140). Springer Nature Singapore. https://doi.org/10.1007/978-981-99-3039-5_11

Chowdhary, S., Iacopini, I., & Battiston, F. (2023). Quantifying human performance in chess. *Scientific Reports, 13*(1), 2113. https://doi.org/10.1038/s41598-023-27735-9

Collie, A., Darby, D., & Maruff, P. (2001). Computerised cognitive assessment of athletes with sports related head injury. *British Journal of Sports Medicine, 35*(5), 297. https://doi.org/10.1136/bjsm.35.5.297

Cotterill, S. T., Loughead, T. M., & Fransen, K. (2022). Athlete leadership development within teams: Current understanding and future directions [review]. *Frontiers in Psychology, 13*. https://www.frontiersin.org/articles/10.3389/fpsyg.2022.820745

Crivelli, D., & Balconi, M. (2022). Neuroassessment in sports: An integrative approach for performance and potential evaluation in athletes [opinion]. *Frontiers in Psychology, 13*. https://www.frontiersin.org/articles/10.3389/fpsyg.2022.747852

Davis, L., Jowett, S., & Tafvelin, S. (2019). Communication strategies: The fuel for quality coach-athlete relationships and athlete satisfaction [original research]. *Frontiers in Psychology, 10*. https://www.frontiersin.org/articles/10.3389/fpsyg.2019.02156

Eric, J. (2023). *Sports analytics: Complete intro to sports data analytics*. Data Rundown. Retrieved 27 October from https://datarundown.com/sports-analytics-intro/

Eys, M., & Kim, J. (2017). *Team building and group cohesion in the context of sport and performance psychology*. Oxford University Press.

Fitzpatrick, W. (2020). *A crash course in basketball analytics*. Samford University. Retrieved 27 October from https://www.samford.edu/sports-analytics/fans/2020/A-Crash-Course-in-Basketball-Analytics

Foster, G., O'Reilly, N., & Naidu, Z. (2021). Playing-side analytics in team sports: Multiple directions, opportunities, and challenges [original research]. *Frontiers in Sports and Active Living, 3*. https://www.frontiersin.org/articles/10.3389/fspor.2021.671601

Foulds, S. J., Hoffmann, S. M., Hinck, K., & Carson, F. (2019). The coach–athlete relationship in strength and conditioning: high performance athletes' perceptions. *Sports, 7*(12).

Fransen, K., McEwan, D., & Sarkar, M. (2020). The impact of identity leadership on team functioning and well-being in team sport: Is psychological safety the missing link? *Psychology of Sport and Exercise, 51*, 101763. https://doi.org/10.1016/j.psychsport.2020.101763

Friend, D. (2022). *The future of sports is in the cloud*. Forbes. Retrieved 27 October from https://www.forbes.com/sites/forbestechcouncil/2022/05/25/the-future-of-sports-is-in-the-cloud/

Fury, M. S., Oh, L. S., & Berkson, E. M. (2022). New opportunities in assessing return to performance in the elite athlete: Unifying sports medicine, data analytics, and sports science. *Arthroscopy, Sports Medicine, and Rehabilitation, 4*(5), e1897–e1902.

Ghosh, I., Ramasamy Ramamurthy, S., Chakma, A., & Roy, N. (2023). Sports analytics review: Artificial intelligence applications, emerging technologies, and algorithmic perspective. *WIREs Data Mining and Knowledge Discovery, 13*(5), e1496. https://doi.org/10.1002/widm.1496

Giorgio, P., & Westcott, K. (2023). *A whole new ball game: Navigating digital transformation in the sports industry*. Deloitte. Retrieved 26 October from https://www2.deloitte.com/us/en/pages/technology-media-and-telecommunications/articles/digital-transformation-and-future-changes-in-sports-industry.html

Grossman, R., Nolan, K., Rosch, Z., Mazer, D., & Salas, E. (2021). The team cohesion-performance relationship: A meta-analysis exploring measurement approaches and the changing team landscape. *Organizational Psychology Review, 12*(2), 181–238. https://doi.org/10.1177/20413866211041157

Hanson, B. (2023). *Top factors for successful recruitment in sport management*. Athlete Assessments. Retrieved 27 October from https://www.athleteassessments.com/6-top-factors-successful-recruitment-in-sport

Houtmeyers, K. C., Jaspers, A., & Figueiredo, P. (2021). Managing the training process in elite sports: From descriptive to prescriptive data analytics. *International Journal of Sports Physiology and Performance, 16*(11), 1719–1723. https://doi.org/10.1123/ijspp.2020-0958

Huyghe, T., Alcaraz, P. E., Calleja-González, J., & Bird, S. P. (2022). The underpinning factors of NBA game-play performance: A systematic review (2001–2020). *The Physician and Sportsmedicine, 50*(2), 94–122. https://doi.org/10.1080/00913847.2021.1896957

Jarvis, D., & Westcott, K. (2021). *The hyperquantified athlete: Technology, measurement, and the business of sports*. Deloitte. Retrieved 24 August from https://www2.deloitte.com/xc/cn/insights/industry/technology/technology-media-and-telecom-predictions/2021/athlete-data-analytics.html

Jones, M., Collier, G., Reinkensmeyer, D. J., DeRuyter, F., Dzivak, J., Zondervan, D., & Morris, J. (2020). Big data analytics and sensor-enhanced activity management to improve effectiveness and efficiency of outpatient medical rehabilitation. *International Journal of Environmental Research and Public Health, 17*(3).

Jowett, S. (2017). Coaching effectiveness: The coach–athlete relationship at its heart. *Current Opinion in Psychology, 16*, 154–158. https://doi.org/10.1016/j.copsyc.2017.05.006

Kao, S.-F., & Tsai, C.-Y. (2016). Transformational leadership and athlete satisfaction: The mediating role of coaching competency. *Journal of Applied Sport Psychology, 28*(4), 469–482. https://doi.org/10.1080/10413200.2016.1187685

Kelly, A. L. (2023). *Talent identification and development in youth soccer: A guide for researchers and practitioners*. Taylor & Francis. https://books.google.com/books?id=32rMEAAAQBAJ

Khanna, A. (2023). *Can AI score big in the future of sports? Five key trends shaping the industry*. Forbes. Retrieved 27 October from https://www.forbes.com/sites/forbestechcouncil/2023/09/27/can-ai-score-big-in-the-future-of-sports-five-key-trends-shaping-the-industry/

Koopmann, T., Faber, I., Baker, J., & Schorer, J. (2020). Assessing technical skills in talented youth athletes: A systematic review. *Sports Medicine, 50*(9), 1593–1611. https://doi.org/10.1007/s40279-020-01299-4

Leano, H. (2021). *Five real-world applications for sports analytics*. Datanami. Retrieved 27 October from https://www.datanami.com/2021/03/18/five-real-world-applications-for-sports-analytics/

Leite, N., Calvo, A. L., Cumming, S., Gonçalves, B., & Calleja-Gonzalez, J. (2021). Editorial: Talent identification and development in sports performance [editorial]. *Frontiers in Sports and Active Living, 3*. https://www.frontiersin.org/articles/10.3389/fspor.2021.729167

Li, B., & Xu, X. (2021). Application of artificial intelligence in basketball sport. *Journal of Education, Health and Sport, 11*(7), 54–67.

Li, Q., Li, X., Chui, K. T., & Arya, V. (2023). Exploring the intersection of athletic psychology and emerging technologies. *International Journal on Semantic Web and Information Systems (IJSWIS), 19*(1), 1–15. https://doi.org/10.4018/IJSWIS.329168

Liu, A., Mahapatra, R. P., & Mayuri, A. V. R. (2023). Hybrid design for sports data visualization using AI and big data analytics. *Complex & Intelligent Systems, 9*(3), 2969–2980. https://doi.org/10.1007/s40747-021-00557-w

Lotfy, C., Bennis Bennani, Y., & Cornet, A. (2023). A meta-analysis of women professional evolution in sports leadership. Une méta-analyse de l'évolution professionnelle des femmes dans le leadership sportif. *Problems and Perspectives in Management, 21*(4), 38–51. https://doi.org/10.21511/ppm.21(4).2023.04

Lyngstad, T. H., Härkönen, J., & Rønneberg, L. T. S. (2020). Nationalistic bias in sport performance evaluations: An example from the ski jumping world cup. *European Journal for Sport and Society, 17*(3), 250–264. https://doi.org/10.1080/16138171.2020.1792628

Mayo Clinic. (2021). *Sports analytics: Identifying at-risk athletes before injuries occur*. Mayo Clinic. Retrieved 27 October from https://www.mayoclinic.org/medical-professionals/orthopedic-surgery/news/sports-analytics-identifying-at-risk-athletes-before-injuries-occur/mac-20517767

McCartney, S., & Fu, N. (2022). Promise versus reality: A systematic review of the ongoing debates in people analytics. *Journal of Organizational Effectiveness: People and Performance, 9*(2), 281–311. https://doi.org/10.1108/JOEPP-01-2021-0013

Mohd Kassim, A. F., Wan Abdullah, W. F. I., Md Japilus, S. J., & Azanuar Yusri, A. S. (2020). *Coach-athlete relationship and coaching effectiveness in team sports athletes*. Enhancing Health and Sports Performance by Design, Singapore.

Montull, L., Slapšinskaitė-Dackevičienė, A., Kiely, J., Hristovski, R., & Balagué, N. (2022). Integrative proposals of sports monitoring: Subjective outperforms objective monitoring. *Sports Medicine—Open, 8*(1), 41. https://doi.org/10.1186/s40798-022-00432-z

Najjar, M.-C. (2023). Legal and ethical issues arising from the application of data analytics and artificial intelligence to traditional sports. *Alb. LJ Sci. & Tech., 33*, 51.

Orlando, A. (2022, 6–8 July 2022). AI for sport in the EU legal framework. *2022 IEEE International Workshop on Sport, Technology and Research (STAR)*

Otto, I. A., Breugem, C. C., Malda, J., & Bredenoord, A. L. (2016). Ethical considerations in the translation of regenerative biofabrication technologies into clinic and society. *Biofabrication, 8*(4), 042001. https://doi.org/10.1088/1758-5090/8/4/042001

Owens, J. G., Rauzi, M. R., Kittelson, A., Graber, J., Bade, M. J., Johnson, J., & Nabhan, D. (2020). How new technology is improving physical therapy. *Current Reviews in Musculoskeletal Medicine, 13*(2), 200–211. https://doi.org/10.1007/s12178-020-09610-6

Patel, D., Shah, D., & Shah, M. (2020). The intertwine of brain and body: A quantitative analysis on how big data influences the system of sports. *Annals of Data Science, 7*(1), 1–16. https://doi.org/10.1007/s40745-019-00239-y

Pavitt, J., Braines, D., & Tomsett, R. (2021). Cognitive analysis in sports: Supporting match analysis and scouting through artificial intelligence. *Applied AI Letters, 2*(1), e21. https://doi.org/10.1002/ail2.21

Pickering, C., & Kiely, J. (2019). The development of a personalised training framework: Implementation of emerging technologies for performance. *Journal of Functional Morphology and Kinesiology, 4*(2).

Pino-Ortega, J., Rojas-Valverde, D., Gómez-Carmona, C. D., & Rico-González, M. (2021). Training design, performance analysis, and talent identification—A systematic review about the most relevant variables through the principal component analysis in soccer, basketball, and rugby. *International Journal of Environmental Research and Public Health, 18*(5).

Plessner, H., & Haar, T. (2006). Sports performance judgments from a social cognitive perspective. *Psychology of Sport and Exercise, 7*(6), 555–575. https://doi.org/10.1016/j.psychsport.2006.03.007

Pykes, K. (2022). *Sports analytics: How different sports use data analytics*. DataCamp. Retrieved 28 August from https://www.datacamp.com/blog/sports-analytics-how-different-sports-use-data-analysis

Randhawa, M. (2023). *AI ethics in people analytics: Balancing innovation and unbiased decision making*. Insight222. Retrieved 27 October from https://www.myhrfuture.com/blog/ai-ethics-in-people-analytics-balancing-innovation-and-unbiased-decision-making

Salter, J., Black, J., Mallett, J., Barrett, S., Towlson, C., Hughes, J. D., & De St Croix, M. (2023). Does biologically categorised training alter the perceived exertion and neuromuscular movement profile of academy soccer players compared to traditional age-group categorisation? *European Journal of Sport Science, 23*(8), 1490–1499. https://doi.org/10.1080/17461391.2022.2117090

Santosh, R. S. (2021). *The Sports Class Thinking towards business success: Unique ideas from sports-field to win in business management*. Notion Press.

Sarlis, V., & Tjortjis, C. (2020). Sports analytics—Evaluation of basketball players and team performance. *Information Systems, 93*, 101562. https://doi.org/10.1016/j.is.2020.101562

Sarmento, H., Anguera, M. T., Pereira, A., & Araújo, D. (2018). Talent identification and development in male football: A systematic review. *Sports Medicine, 48*(4), 907–931. https://doi.org/10.1007/s40279-017-0851-7

Seçkin, A. Ç., Ateş, B., & Seçkin, M. (2023). Review on wearable technology in sports: Concepts, challenges and opportunities. *Applied Sciences, 13*(18).

Seidel-Marzi, O., & Ragert, P. (2020). Neurodiagnostics in sports: Investigating the Athlete's brain to augment performance and sport-specific skills [perspective]. *Frontiers in Human Neuroscience, 14*. https://www.frontiersin.org/articles/10.3389/fnhum.2020.00133

Sharma, D., Singh Aujla, G., & Bajaj, R. (2023). Evolution from ancient medication to human centered Healthcare 4.0: A review on health care recommender systems. *International Journal of Communication Systems, 36*(12), e4058. https://doi.org/10.1002/dac.4058

Shuffler, M. L., Diazgranados, D., Maynard, M. T., & Salas, E. (2018). Developing, sustaining, and maximizing team effectiveness: An integrative, dynamic perspective of team development interventions. *Academy of Management Annals, 12*(2), 688–724. https://doi.org/10.5465/annals.2016.0045

Smith, R. E., & Smoll, F. L. (2017). *Coaching behavior and effectiveness in sport and exercise psychology*. Oxford University Press.

Stephenson, S. D., Kocan, J. W., Vinod, A. V., Kluczynski, M. A., & Bisson, L. J. (2021). A comprehensive summary of systematic reviews on sports injury prevention strategies. *Orthopaedic Journal of Sports Medicine, 9*(10), 23259671211035776. https://doi.org/10.1177/23259671211035776

Strength, N. N., Conditioning, A., French, D., & Ronda, L. T. (2021). *NSCA's essentials of sport science*. Human Kinetics. https://books.google.com/books?id=1GIXEAAAQBAJ

Suri, N., & Lakhanpal, P. (2022). People analytics enabling HR strategic partnership: A review. *South Asian Journal of Human Resources Management, 23220937221119599*. https://doi.org/10.1177/23220937221119599

Tee, J. C., McLaren, S. J., & Jones, B. (2020). Sports injury prevention is complex: We need to invest in better processes, not singular solutions. *Sports Medicine, 50*(4), 689–702. https://doi.org/10.1007/s40279-019-01232-4

Thomas, A. (2023). Football Analytics: A Real-World Case Examples. Linkedin. Retrieved 27 October from https://www.linkedin.com/pulse/football-analytics-real-world-case-examples-alokh-thomas/

Thornton, H. R., Delaney, J. A., Duthie, G. M., & Dascombe, B. J. (2019). Developing athlete monitoring systems in team sports: Data analysis and visualization. *International Journal of Sports Physiology and Performance, 14*(6), 698–705. https://doi.org/10.1123/ijspp.2018-0169

Till, K., & Baker, J. (2020). Challenges and [possible] solutions to optimizing talent identification and development in sport [hypothesis and theory]. *Frontiers in Psychology, 11*. https://www.frontiersin.org/articles/10.3389/fpsyg.2020.00664

Torres-Ronda, L., Beanland, E., Whitehead, S., Sweeting, A., & Clubb, J. (2022). Tracking systems in team sports: A narrative review of applications of the data and sport specific analysis. *Sports Medicine—Open, 8*(1), 15. https://doi.org/10.1186/s40798-022-00408-z

Tursunbayeva, A., Di Lauro, S., & Pagliari, C. (2018). People analytics—A scoping review of conceptual boundaries and value propositions. *International Journal of Information Management, 43*, 224–247. https://doi.org/10.1016/j.ijinfomgt.2018.08.002

Vaeyens, R., Lenoir, M., Williams, A. M., & Philippaerts, R. M. (2008). Talent identification and development programmes in sport. *Sports Medicine, 38*(9), 703–714. https://doi.org/10.2165/00007256-200838090-00001

Van Eetvelde, H., Mendonça, L. D., Ley, C., Seil, R., & Tischer, T. (2021). Machine learning methods in sport injury prediction and prevention: A systematic review. *Journal of Experimental Orthopaedics, 8*(1), 27. https://doi.org/10.1186/s40634-021-00346-x

Vella, S. A., Mayland, E., Schweickle, M. J., Sutcliffe, J. T., McEwan, D., & Swann, C. (2022). Psychological safety in sport: A systematic review and concept analysis. *International Review of Sport and Exercise Psychology, 1-24*. https://doi.org/10.1080/1750984x.2022.2028306

West, S. W., Clubb, J., Torres-Ronda, L., Howells, D., Leng, E., Vescovi, J. D., Carmody, S., Posthumus, M., Dalen-Lorentsen, T., & Windt, J. (2020). More than a metric: How training load is used in elite sport for athlete management. *International Journal of Sports Medicine, 42*(04), 300–306. https://doi.org/10.1055/a-1268-8791

Zadeh, A., Taylor, D., Bertsos, M., Tillman, T., Nosoudi, N., & Bruce, S. (2021). Predicting sports injuries with wearable technology and data analysis. *Information Systems Frontiers, 23*(4), 1023–1037. https://doi.org/10.1007/s10796-020-10018-3

Identifying Factors Influencing Fan Engagement: A Pilot Study on Chennai Superkings

S. Shyni Carmel Mary, Joe Arun, and M. J. Xavier

1 Introduction

Data analytics is crucial for solving business problems, aiding in efficient decision-making, and contributing to revenue generation (Mamo et al., 2022). It plays a pivotal role in enhancing revenue alignment with business objectives. Sports represent a nation's culture and unity, instilling values like determination and discipline Gupta & Divekar (2022). Sports events foster sportsmanship and gauge public affinity for a nation's sports endeavors. Data plays a multifaceted role in sports, improving aspects like planning, organization, and budgeting, with accurate data leading to favorable outcomes. Conventional sports management carries biases, to be explored in forthcoming chapters. Data analytics introduces techniques to address these challenges.

Indian sports embody athleticism, cultural heritage, and the nation's identity (Glebova et al., 2023), transcending recreation and reflecting diversity. Cricket is a unifying force in India Noorbhai (2022), representing the hopes and dreams of millions. Beyond cricket, sports like hockey, badminton, chess, and weightlifting contribute to our sporting legacy. Indian sports foster camaraderie, bridge generational gaps, and unite diverse traditions. Athletes inspire with their determination and the spirit of "never give up," celebrating our heritage and perseverance. Indian sports are more than scores and medals; they symbolize the collective undying spirit and serve as a canvas for the dreams of the fans. The cheer by the fans ends up motivating the players for larger triumphs.

The Indian Premier League (IPL) is not just a sport but it has gone into the psyche of every Indian cricket fan. The IPL tournament was started in 2008 with eight teams from eight Indian cities. It chose the T20 format, as opposed to the

S. S. C. Mary (✉) · J. Arun · M. J. Xavier
Loyola Institute of Business Administration (LIBA), Chennai, Tamil Nadu, India
e-mail: Shyni.pragasam@liba.edu; xavier.mj@liba.edu

A. Mansurali et al. (eds.), *Sports Analytics*,
https://doi.org/10.1007/978-3-031-63573-1_6

longer version that is played for 5 days. As such, this cricket tournament generates a large amount of data, covering players, games, every over, and literally every ball. This treasure trove is ideally suited for sports analytics, providing insights into the performance of each team and every player. The data can be used for strategizing matches and improving individual performance. By combining fan engagement data with game data, additional insights can be gained.

In this article, we explore the dynamic realm of sports analytics and its pivotal role in fostering efficient fan engagement, leading to a substantial increase in sports revenue (Anuraj et al., 2023). We focus on the interplay between fan involvement and improvement in the brand, which will increase the revenue of the sports, particularly within IPL teams. Our detailed case study is focused on the Chennai Super Kings (CSK), analyzing fan sentiments and the data from a survey among CSK supporters and the public on live interactions.

2 Literature Review

The literature is to highlight the growing importance of analytics in sports. Sports analytics and its origin come into focus, underscoring its significance and applications across various sectors. It delineates the diverse goals and interests of various stakeholders, encompassing professional sports clubs, media companies, IT firms, academics, sports enthusiasts, and computer scientists. Sports analytics is deriving insights from big data Kaur et al. (2021). A range of diverse analytics techniques has been harnessed across these studies.

Typically, analytics in sports has been used for performance analysis, player recruitment, predicting outcomes, injury prevention, player behavior analysis, and others. While these studies have extensively employed a variety of analytics techniques to delve into the intricacies of sports performance, it is equally important to acknowledge the role of fan engagement in shaping the landscape of sports. Fan engagement, a crucial aspect of the sports ecosystem, influences various dimensions, from team loyalty to revenue generation. Just as analytics contribute to a comprehensive understanding of sports dynamics, understanding fan behaviors and preferences through engagement strategies is imperative for creating an immersive and interactive sports experience. Thus, integrating advanced analytics with fan engagement initiatives can potentially enhance not only on-field performance analysis but also off-field experiences, contributing to a holistic approach in sports management and research.

The studies analyzed in this summary utilize a range of methodologies to investigate and enhance fan engagement in sports. These approaches include (i) interactive visualizations and technology integration to create immersive experiences, (ii) data analysis and sentiment analysis to uncover engagement trends and emotions, (iii) systematic analytics strategies for data-driven decision-making, (iv) examination of fandom behavior and identity's role in engagement, (v) exploration of interactive features like bullet-screens, and (vi) the study of self-categorization's impact on engagement. Together, these studies contribute to a holistic understanding of factors influencing fan engagement in sports, as given in Table 1.

Table 1 Different analytics methods used for the sports problem

Article title	Purpose	Methods used for fan engagement
Fan Engagement in Sports Data Analytics, Naraine & Bakhsh (2022)	To enhance basketball fans' engagement through interactive visualizations	User-centered design for interactive embedded visualizations
Sport Technology Innovations, Wu et al. (2022)	To integrate technology for enhanced fan experiences in sports events	Development of technology systems and integration in technology-enhanced broadcasting system (TEBS)
Social Media Engagement in Sport Context, Jakar & Carr (2023)	To analyze social media engagement for professional sports teams	Data analysis, correlation, and exploration of social media engagement
Utilizing Data in Sports Business, Bedeley et al. (2016)	To develop a systematic analytics strategy for revenue generation in the sports business	Data-driven decision-making strategies, qualitative case study approach
International Discourse of Chinese Super League, Laurell et al. (2021)	To examine which actors are driving the current developments, framing, and engagement of the Chinese super league discourse on social media	Analysis of international discourse and social media engagement (SMA)
Ultra-High-Definition Services in Sports, Kennedy et al. (2021)	To explore the adoption and impact of ultra-high-definition services on sports fan engagement	Exploration of evolving viewing behaviors and social analytics
Sentiment Analysis of Sports Fans on Twitter, Hoeber et al. (2016)	To determine sports fans' sentiments and opinions on twitter and their correlation with engagement	Sentiment analysis of twitter data
Systematic Analytics Strategy in Sports, Weimar et al. (2021)	To establish a systematic analytics strategy for enhancing revenue and performance in the sports business. Used statistical regression to show the significance of core drives for fan engagement	Data-driven decision-making strategies, case studies
Sport Fans' Hashtags Usage, Kim et al. (2021)	To investigate how sports fans use hashtags for information sharing on social media and sentiment analysis	Text analysis, sentiment analysis, and algorithm comparison
Case Study of Foot Locker Elite Classic, Scibilia & Hutchins (2011)	To create an engaging event for basketball fans through real-time analytics and fan culture understanding	Collaborative efforts, real-time analytics, and event creation
Sports fans' blogs and media interaction, McCarthy (2013)	To understand the behavior of fan sports bloggers and their interaction with sports media	Qualitative analysis, thematic analysis, interviews

(continued)

Table 1 (continued)

Article title	Purpose	Methods used for fan engagement
Strategy in maximising fan's engagement Anderson & Xu, (2019)	To investigate on how the fan engagement is maximised due to live tweeting messaging strategy	Data analysis, examination of live message strategy
University Student's fandom behavior, Melnick & Wann (2011)	To explore sports fandom patterns among university students and compare them with international samples	Survey, statistical analysis, and comparison with international samples
Fan Social-Personal Identity Salience, Shuv-Ami & Toder Alon (2023)	To develop a scale measuring fans' personal and social identity salience and its impact on engagement	Scale development, structural equation model, analysis
Bullet-Screen Function and Sports, Fans Li et al. (2023)	To investigate how fans engage with the bullet-screen feature during live sports streaming for engagement	Thematic analysis, interviews, qualitative analysis
Self-Categorization of Fans in Sports, Shuv-Ami & Toder Alon (2023)	To explore how fans self-categorize based on social identity and its impact on fan engagement	Self-categorization study, analysis of motives and identity impact

The way we use numbers and information in sports is expected to get better in the future. This is because new technologies, machines that learn, and science about data are improving a lot. When we use these improvements, we can guess what might happen in sports more accurately, understand things better, and make smarter choices in studying sports (Morgulev et al, 2018). By looking at things like social media and how fans react, we can learn what makes them happy.

2.1 Understanding the Factors in Fans Engagement

Fan engagement in sports is influenced by a range of factors that contribute to fans' excitement and involvement. These factors include the geographic location of fans, with local fans often showing a stronger connection to their teams. The type of game also matters as significant matches like playoffs or finals tend to generate heightened engagement. Star players play a key role, as their performances evoke pride and enthusiasm among fans. The timing of games influences engagement, with convenient timings attracting more viewers and discussions. The impact of social media is undeniable, as fans share posts, comments, and pictures that create a vibrant online community. Moments of triumph, such as critical catches or players reaching 50 s and 100 s, trigger surges in excitement, uniting fans in celebration. Exclusive offers and interactive content further enhance engagement, while shared experiences and fan communities foster a sense of belonging. Cultural influences, feedback integration, and even long-term engagement trends all play their part in shaping the dynamics of fan engagement. Understanding and harnessing these factors empower sports teams to

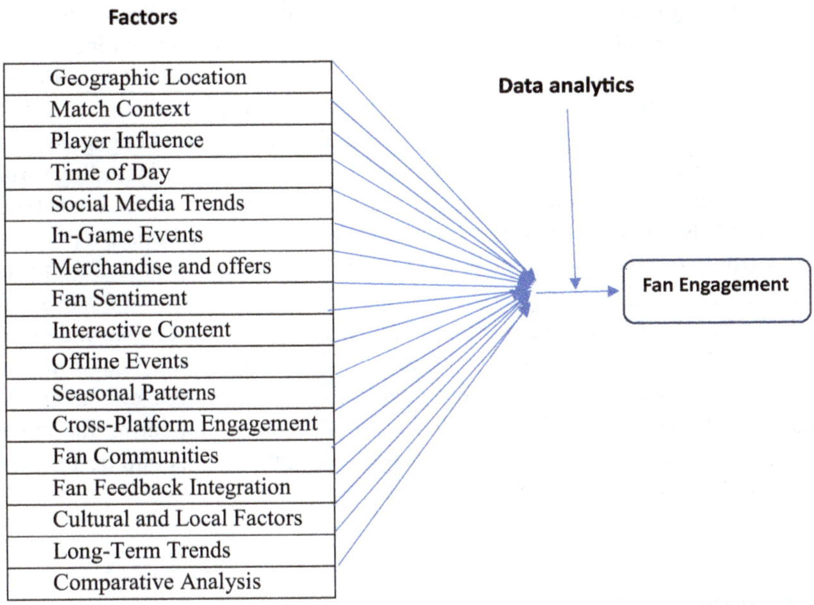

Fig. 1 Factors considered for the fan engagement. (Data Source: Author)

create compelling and meaningful connections with their passionate fan base. A comprehensive fan engagement analysis involves a holistic approach that considers various factors and their interactions Sikka et al. (2019). The goal is to continuously refine and tailor engagement strategies based on data-driven insights. The factors considered for the fan engagement research as shown in Fig. 1 are based on Huettermann et al. (2022), Yoshida et al. (2014), and Muniz & Flamand (2022).

Geographic Location: Analyze fan engagement trends based on the geographic location of fans. Understand whether engagement patterns vary between local fans and those from different regions. *Match Context*: Consider the context of the match, such as whether it's a home game or an away game, the significance of the match (playoffs, finals), and the opponent team's popularity. *Player Influence*: Analyze the impact of star players on fan engagement. Do fans engage more when certain players are performing exceptionally well? *Time of Day*: Explore whether engagement varies depending on the time of day when matches are held. Consider time zone differences for international audiences. *Social Media Trends*: Monitor social media trends and current events that may affect fan engagement. Leveraging trending topics can enhance engagement. *In-Game Events*: Analyze how in-game events (e.g., goals, wickets, home runs) impact fan engagement. Consider real-time engagement spikes during critical moments. *Merchandise and Offers*: Assess how merchandise promotions, discounts, and exclusive offers impact fan engagement and purchase behavior. *Fan Sentiment*: Utilize sentiment analysis to gauge fan reactions and emotions related to match outcomes, team performance, and other related factors.

Interactive Content is evaluating the effectiveness of interactive content, such as polls, quizzes, and challenges, in boosting fan engagement. *Offline Events* include analysis of offline events like fan meet-ups, watch parties, and stadium experiences in the overall engagement assessment (Piché & Naraine., 2022). *Seasonal Patterns* is looking for seasonal patterns in fan engagement. Does engagement increase during specific times of the year (e.g., playoffs, festive seasons)? *Cross-Platform Engagement* is a study of how fans engage across various platforms (e.g., website, mobile app, social media, email) and whether there are consistent engagement patterns. *Fan Communities* is analyzing fan community interactions, including forums, social media groups, and online discussions. Do these communities contribute to engagement? (Naraine & Bakhsh., 2022). *Fan Feedback Integration* incorporates feedback from fan surveys, comments, and reviews to understand areas for improvement in fan engagement strategies (Santos et al., 2019). *Cultural and Local Factors* consider cultural nuances and local customs that may influence fan engagement behaviors and preferences. *Long-Term Trends* is analyzing engagement trends over multiple seasons to identify long-term patterns and assess the effectiveness of engagement strategies over time. *Comparative Analysis* is comparing fan engagement metrics with other sports teams or leagues to gain insights into best practices and potential areas for improvement.

3 Methodology

After making a clear definition of the problem, the research methodology was devised. The research methodology is schematically shown in Fig. 2. Using a detailed literature review of literature review, various aspects of fan engagement

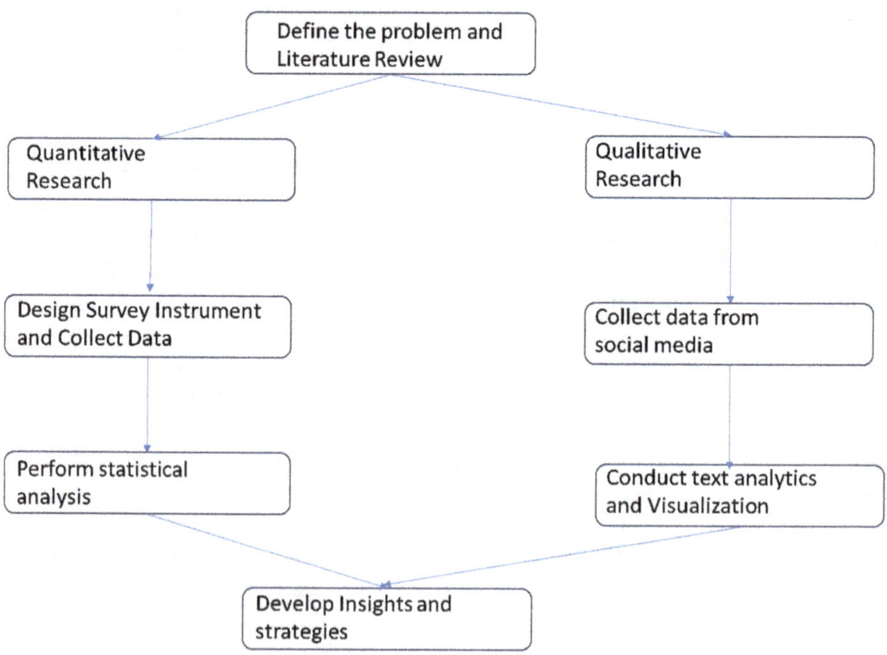

Fig. 2 Steps for deriving insight (factors) from the data of CSK fan engagement. (Data source: Authors)

were identified (see Fig. 1). Data collection was carried out in two phases. In phase one, quantitative data was collected using a survey instrument developed based on the aspects of live interactions of the fans identified through a literature search. In phase two, the comments and reactions of the fans were scrapped from the social media and official websites.

The data analysis for the survey data was carried out using SPSS. The qualitative data from social media was cleaned and subjected to sentiment analysis using Python. The insights derived from both analyses were compared to arrive at strategies for better fan engagement and revenue enhancement.

4 Discussions: A Case Study on Chennai Super Kings (IPL)

Indian Premier League is a professional Twenty20 cricket league in India (IPL—Indian Premier League, Hindustan Times). It was established in 2008 by the Board of Control for Cricket in India (BCCI) with the goal of creating a high-energy, globally recognized cricket competition that brings together top talent from around the world. There were ten teams in Indian Premier League cricket. The IPL is known for its entertainment quotient, with cheerleaders, celebrity performances, and fan engagement initiatives that contribute to the festive atmosphere during matches (IPL official website). The IPL matches are typically held over a couple of months, with multiple matches taking place on most days, making it a cricketing extravaganza for fans (Wikipedia page: Chennai Super Kings in 2023). In 2023, the 16th season of the IPL, a franchise Twenty20 cricket league in India, was organized by the Board of Control for Cricket in India.

The IPL tournament is not only a successful game but also generates a substantial amount of revenue through sponsorships, broadcasting rights, and advertisements. It has helped many players to get global visibility and fame. The revenue generation is made possible through its massive fan support and following not only from India but also from many other countries such as the Gulf, Australia, and UK. It has also become a major economic activity for India generating jobs and developing ancillary industries.

Of the different teams that participate in the IPL, the Chennai Super Kings has emerged as the most popular team with a very high fan following. It started its journey in a humble way in 2008. The team got instant recognition and it was led by the captain of the then Indian cricket team MS Dhoni. The team was able to ride on the image of Captain and quickly became a major force to reckon with. The team reached the finals in their debut year itself, but narrowly missed the trophy. However, by 2010 it bounced back and won the IPL Trophy. The winning streak continued in 2011 also. The leadership of Dhoni and the selection of players helped the team to attain an iconic position. It managed to attract a huge fan following which was followed by additional revenue generated by brand merchandise like T shirts, caps, etc. The period from 2012 to 2017 was one of lull, as the team faced many challenges such as injuries, team controversies, and suspensions. However, the fan following for the CSK didn't dwindle to such low levels.

The year 2018 saw the CSK team come back with a thumping victory. This made the fans of CSK come back invigorated and with greater passion. Dhoni continued with the team from 2018. In 2019, they reached the runner-up position and in 2020 they reached the playoffs stage but could not make it to the finals.

The 2021 season saw CSK focusing on rebuilding their squad with a mix of experienced players and promising youngsters. While they didn't lift the trophy, the team displayed promising performances, with players like Ruturaj Gaikwad and Sam Curran making their mark. Though the Chennai Super Kings had lifted the trophy four times, in 2010, 2011, 2018, and 2021, the team got pushed to the ninth position in 2022. They also went through a leadership change to Ravindra Jadeja and back again to MS Dhoni.

In 2023, CSK was the first team to seal a spot in the IPL 2023 final and is an *iconic IPL brand*. They appeared in the summit clash of the league for a record tenth time and lifted the title once again.

As per the report by Houlihan Lokey, the brand value of the IPL has reached an impressive USD 3 billion. This substantial increase underscores the league's growing influence and popularity. Over the years, the IPL has transformed from a fledgling concept to a sporting powerhouse with a remarkable brand presence. This achievement reflects the league's ability to *captivate audiences*, *attract sponsors*, and secure a prominent place in the global cricket landscape. The surge in brand value signifies the sustained success and impact of the IPL on the sports industry. "The IPL's media rights have grown at a CAGR of 18% from 2008 to 2023. CSK is the ranked as no 1, both in brand ranking and business enterprise value ranking (CSK Most Valuable Brand in IPL. (n.d.). Sportskeeda). An Iconic brand, CSK were valued at USD 146 million in 2022, courtesy the massive popularity of MS Dhoni" (*Times of India*, 2023; Updated: Jul 10, 2023, 18:18 IST). Throughout their IPL journey from 2008 to 2021, Chennai Super Kings maintained their status as a team with a strong fan base, consistent performances, and a winning mentality (How to Boost Fan Engagement in Modern Sports—2022). With several titles and a legacy of success, CSK has cemented its position as one of the most iconic franchises in the history of the Indian Premier League. The Brand Finance IPL report (2017, 2018) discusses the interconnection between brand value and revenue increase.

Teams generate revenue in various ways, including official and dedicated sponsors, broadcasting rights, ticket sales, merchandise, and player transfers. Media rights and sponsorships make up a significant part of revenue, tied to the team's fan base. Ultimately, fan numbers contribute a great deal to revenue generation.

There are a few questions raised, such as: What influences an IPL team's fan following? Is a team's fan engagement linked to players or the Team? Does a team's celebrity association affect its fan base? Is there a connection between city/state population and a team's fan strength? Is fan engagement improving the brand revenue in IPL? To improve the strategy used for the team in the IPL, it is important to increase the number of fans and their engagement, which will impact the brand and revenue. Fan following and engagement can be studied by measuring social media posts, likes, and follows.

This study looks at the case of Chennai Super Kings and has come up with recommendations specific to CSK, which has the highest fan loyalty (Chennai Super Kings has highest fan loyalty: Rediffusion Red Lab). The model can be potentially extended to other IPL teams by incorporating the demographics of different teams.

4.1 Data Used in the Study

Survey data was collected from the fans to understand their level of engagement on different factors listed in Fig. 2. Out of a total of 450 responses collected, 20 were incomplete and not considered for the model development. Additionally, comments by the fans on social media were scraped from Twitter, Facebook, Instagram, and the CSK website. This data was subjected to NLP.

4.2 Statistical Analysis to Identify the Key Factors Responsible for fan Engagement

A factor analysis Kim & Mueller (1978) was conducted on a set of statements that were rated by fans. These statements are interrelated, and Bartlett's test has verified that they are suitable for conducting factor analysis. This statistical test affirms that the relationships between these statements are significant enough to extract meaningful factors. Through factor analysis, this study delves into the underlying structure of these statements, aiming to unveil any hidden patterns or dimensions that might exist based on how fans have assigned their ratings. This approach allows you to understand the connections between these statements better and potentially group them into categories that share common themes or concepts. Factor analysis of different statements rated by fans was carried out.

The statements used in factor analysis in Table 2 are correlated with each other and the Bartlett's test confirms that the same can be used for factor analysis.

Four factors were extracted based on the Scree plot shown in Fig. 3.

Table 2 KMO and Bartlett's test for factor analysis

KMO and Bartlett's test		
Kaiser-Meyer-Olkin measure of sampling adequacy		0.653
Bartlett's test of sphericity	Approx. Chi-Square	832.039
	df	300
	Sig.	0.000

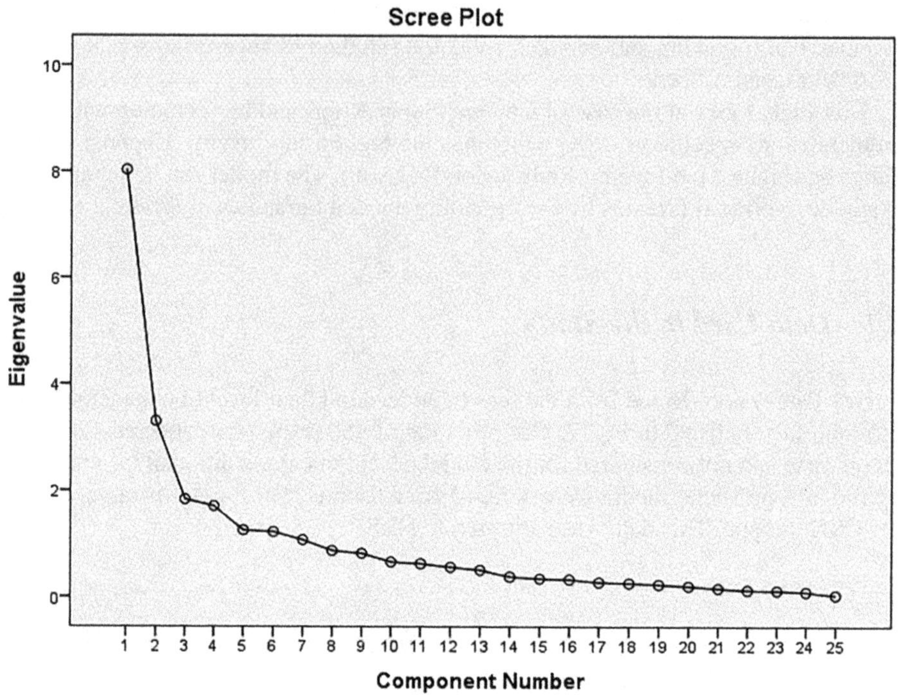

Fig. 3 Scree plot for selecting factors

After meticulous scree plot analysis, a pivotal juncture emerged that distinctly guided our factor selection process. As the factors were plotted in descending order of eigenvalues, a discernible shift in the plot's trajectory occurred beyond the fourth factor. This shift, marked by a conspicuous change in the curve's steepness, signified a point of transition wherein the eigenvalues ceased to exhibit substantial variance. Consequently, adhering to the principle of retaining factors that contribute significantly to the underlying data's variability, we decided to proceed with a selection of four factors in Fig. 3. This selection aligns seamlessly with the inflection point observed in the scree plot, denoting a pragmatic choice that encapsulates the essence of the data while mitigating the inclusion of extraneous and less impactful factors beyond this critical threshold.

The provided data presents the results of a principal component analysis (PCA) performed on the factor analysis. This technique condenses the information in a dataset into a smaller set of components that explain most of the variance. In this case, the analysis has identified several features or factors and their corresponding eigenvalues, variance percentages, and cumulative percentages. The initial eigenvalues indicate the amount of variance explained by each component. The extraction and rotation sums of squared loadings show how much of the total variance each component contributes. This breakdown helps you understand the significance of each factor in explaining the observed variability within the statements rated by fans. The cumulative percentage demonstrates how much of the total variance is explained as you move through the components. For instance, the first component explains 32.124% of the variance, and when combined with the second component, they account for 45.325% of the variance. The cumulative variance explained by the first three factors stands at 52.660%, and the fourth one is at 59.492%, as seen in Table 3. Overall, this PCA helps you identify the most influential factors that capture the essence of the rated statements and how they relate to each other in terms of variance explained. The four factors account for 59.6% of the variance.

The factor loading matrix is reproduced in Table 4.

"Live Interactions in Discussions" indicates that fans who actively participate in online fan communities and discussions, sharing their views and thoughts, have a significant factor loading of 0.809. This component signifies the importance of

Table 3 Total variance of factors

Total variance explained									
	Initial eigenvalues			Extraction sums of squared loadings			Rotation sums of squared loadings		
Component	Total	% of variance	Cumulative %	Total	% of variance	Cumulative %	Total	% of variance	Cumulative %
1	8.031	32.124	32.124	8.031	32.124	32.124	6.188	24.751	24.751
2	3.300	13.201	45.325	3.300	13.201	45.325	4.744	18.977	43.729
3	1.834	7.335	52.660	1.834	7.335	52.660	2.007	8.027	51.756
4	1.708	6.831	59.492	1.708	6.831	59.492	1.934	7.736	59.492
5	1.249	4.997	64.489						
6	1.218	4.870	69.359						
7	1.068	4.272	73.631						
8	0.867	3.467	77.098						
9	0.816	3.262	80.360						
10	0.655	2.621	82.981						
11	0.622	2.489	85.470						
12	0.559	2.236	87.706						
13	0.505	2.019	89.725						
14	0.379	1.516	91.240						
15	0.344	1.376	92.617						
16	0.328	1.313	93.929						
17	0.276	1.104	95.034						
18	0.260	1.041	96.075						
19	0.235	0.941	97.015						
20	0.202	0.810	97.825						
21	0.163	0.650	98.475						
22	0.130	0.521	98.997						
23	0.122	0.489	99.485						
24	0.096	0.384	99.869						
25	0.033	0.131	100.000						

Extraction method: Principal component analysis

Table 4 Rotated component matrix for each factor

Rotated component matrix[a]				
	Component			
	1	2	3	4
21. Do you participate in online fan communities or groups where discussions about the team take place?	0.809	0.084	0.183	−0.223
22. Have you ever provided feedback to the team regarding fan engagement initiatives?	0.756	0.025	−0.006	−0.370
12. Have you ever participated in live interactions (polls, chats) during important in-game events?	0.748	0.127	−0.132	−0.197
13. Have exclusive merchandise offers or discounts influenced your engagement with the team's online store?	0.743	0.138	−0.211	−0.113
17. How 4 are you to participate in interactive content during match broadcasts?	0.739	0.085	0.097	0.250
13. Do special merchandise releases or promotions impact your overall engagement with the team?	0.732	0.077	0.185	0.096
18. Have you ever attended offline fan meet-ups, watch parties, or stadium events organized by the team?	0.715	0.005	0.178	−0.003
16. Have interactive elements like polls, quizzes, or challenges enhanced your engagement experience?	0.715	0.207	0.375	0.115
9. Have you ever engaged with the team's content related to current trending topics or viral challenges?	0.687	0.500	0.020	0.058
15. Have you ever shared your emotions or reactions related to the team's performance on social media?	0.604	0.354	0.152	−0.052
10. Do you feel more motivated to engage when the team participates in social media trends?	0.587	0.312	0.075	0.444
3. How excited are you about matches that have playoff implications or are part of the finals?	0.211	0.851	0.083	0.009
6. Have exceptional individual performances by players ever led you to engage more with the team's content?	0.224	0.847	0.248	−0.086
27. How many years have you been a fan of the Chennai super kings (CSK)?	−0.109	0.750	−0.102	−0.028
4. Do you engage differently with the team during matches against rival teams or iconic opponents?	0.271	0.707	−0.077	−0.148
11. How do goals (runs/wickets) and key moments in matches influence your level of engagement?	0.272	0.678	−0.069	0.181
14. How do match outcomes, especially close wins or losses, affect your emotional engagement with the team?	0.119	0.659	0.373	0.150
2. How does your location impact your engagement with the team's events and content?	0.048	0.533	0.007	0.334
5. Which players' performances do you closely follow and engage with during matches?	−0.038	0.523	0.266	−0.331
24. Have you noticed any changes in your engagement patterns with the team over multiple seasons?	0.174	0.134	0.659	0.125
8. How do time zone differences impact your ability to engage with live matches?	0.126	0.126	−0.643	0.120

(continued)

Table 4 (continued)

Rotated component matrix[a]	Component			
	1	2	3	4
23. Are there any cultural or local factors that influence how you engage with the team's content and events?	0.234	0.132	0.525	−0.154
1. Which city or region do you primarily follow the Chennai super kings from?	−0.122	−0.077	−0.037	0.775
19. Do you find yourself more engaged during certain times of the year, such as playoffs or festive seasons?	0.176	0.401	0.477	0.517
25. Are there any sports teams or leagues that you also engage with, and how does your engagement compare between them?	−0.384	0.016	−0.122	0.425

Extraction method: Principal component analysis
Rotation method: Varimax with Kaiser normalization
[a]Rotation converged in 5 iterations

real-time interactions during both on and off-match discussions. "Feedback and Initiatives" reveals that providing feedback to the team about fan engagement initiatives has a high factor loading of 0.756. This suggests that fan involvement in improving engagement strategies is notable. Engagement During Important Event" implies that engaging fans in live interactions like polls and chats during crucial in-game events holds a substantial factor loading of 0.748. This component highlights the impact of real-time interactions during critical moments. "Merchandise Offers and Engagement" signifies that exclusive merchandise offers or discounts influencing engagement with the team's online store show a strong factor loading of 0.743. This indicates the connection between special offers and increased fan engagement. "Interactive Content Participation" denotes that fans inclined to participate in interactive content during match broadcasts have a notable factor loading of 0.739. This component underscores the significance of interactive elements during match viewing. "Merchandise Impact on Engagement" indicates that special merchandise releases or promotions impacting overall engagement with the team have a considerable factor loading of 0.732. This points out the influence of merchandise-related activities on fan engagement.

"Offline Fan Engagements" implies that attending offline fan meet-ups, watch parties, or stadium events organized by the team is represented by a factor loading of 0.715. This suggests the value of in-person fan gatherings. "Interactive Elements and Engagement" denotes that interactive elements like polls, quizzes, or challenges enhancing engagement experience share the same factor loading of 0.715. This reaffirms the importance of interactive content. "Trending Topics and Viral Challenges" reveals that engaging with the team's content related to current trending topics or viral challenges is linked to a factor loading of 0.687. This reflects the connection between contemporary trends and fan involvement. "Social Media Sharing" points out that sharing emotions or reactions about the team's performance on social media is indicated by a factor loading of 0.604. This emphasizes the role of social media in emotional expression. "Team Spirit and Emotional Connection" denotes that the excitement about matches with playoff implications or finals has a significant factor loading of 0.851. This signifies the emotional connection fans have with these crucial games. "Individual Performances and Engagement" implies that an exceptional individual player performance leading to increased engagement with the team's content holds a strong factor loading of 0.847. This highlights the impact of player excellence on fan interaction.

"Fan Loyalty Duration" shows that the number of years a fan has supported the Chennai Super Kings (CSK) has a notable factor loading of 0.750. This reflects the importance of long-term fan commitment. "Rivalry Matches Engagement" reveals that engaging differently with the team during matches against rival teams or iconic opponents is indicated by a factor loading of 0.707. This underscores the heightened engagement during competitive matchups. "Goals and Key Moments" denotes that the influence of goals (runs/wickets) and key moments in matches on fan engagement is captured by a factor loading of 0.678. This emphasizes the connection between game-changing events and involvement. "Emotional Impact of Match Outcomes" shows that how match outcomes, especially close wins or losses, affect emotional engagement with the team holds a factor loading of 0.659. This highlights the emotional rollercoaster fans experience based on results. "Location's Impact" signifies that the impact of location on engagement with team events and content is represented by a factor loading of 0.533. This shows how geographical factors influence fan involvement. "Player Engagement" reveals that closely following and engaging with specific players' performances during matches has a factor loading of 0.523. This points out the connection between player involvement and fan engagement. "Engagement Pattern Over Time" indicates that noticing changes in engagement patterns with the team over multiple seasons shares a significant factor loading of 0.659. This suggests the recognition of evolving fan engagement trends. "Time Zone Challenges" denotes that the impact of time zone differences on the ability to engage with live matches has a factor loading of −0.643. This underlines the challenges posed by time differences.

"Cultural and Local Factors" reveals that cultural or local factors influencing engagement with the team's content and events are represented by a factor loading of 0.525. This highlights the role of cultural context. "City and Region Engagement" shows that the city or region primarily followed for the Chennai Super Kings has a strong factor loading of 0.775. This signifies the strong bond fans have with their geographical location. "Seasonal Engagement" indicates that being more engaged during specific times of the year, like playoffs or festive seasons, is indicated by a factor loading of 0.517. This shows the influence of seasonal factors on involvement. "Comparative Engagement" reveals that engaging with other sports teams or leagues and comparing engagement between them is linked to a factor loading of 0.425. This suggests assessing involvement across different sports. Based on consultations with the experts, four factors were labeled, as given in Table 5.

These factor loadings and labels reveal the various aspects of fan engagement that contribute to the overall understanding of fan behavior and interaction with the Chennai Super Kings.

Table 5 Four labels for fan engagement

Variables	Factor loading	Factor label
21. Do you participate in online fan communities or groups where discussions about the team take place?	0.809	Live interactions (online and offline)
22. Have you ever provided feedback to the team regarding fan engagement initiatives?	0.756	
12. Have you ever participated in live interactions (polls, chats) during important in-game events?	0.748	
13. Have exclusive merchandise offers or discounts influenced your engagement with the team's online store?	0.743	
17. How likely are you to participate in interactive content during match broadcasts?	0.739	
13. Do special merchandise releases or promotions impact your overall engagement with the team?	0.732	
18. Have you ever attended offline fan meet-ups, watch parties, or stadium events organized by the team?	0.715	
16. Have interactive elements like polls, quizzes, or challenges enhanced your engagement experience?	0.715	
9. Have you ever engaged with the team's content related to current trending topics or viral challenges?	0.687	
15. Have you ever shared your emotions or reactions related to the team's performance on social media?	0.604	
10. Do you feel more motivated to engage when the team participates in social media trends?	0.587	
3. How excited are you about matches that have playoff implications or are part of the finals?	0.851	Team spirit/emotionally connected with the team
6. Have exceptional individual performances by players ever led you to engage more with the team's content?	0.847	
27. How many years have you been a fan of the Chennai Super Kings (CSK)?	0.750	
4. Do you engage differently with the team during matches against rival teams or iconic opponents?	0.707	
11. How do goals (runs/wickets) and key moments in matches influence your level of engagement?	0.678	
14. How do match outcomes, especially close wins or losses, affect your emotional engagement with the team?	0.659	
2. How does your location impact your engagement with the team's events and content?	0.533	
5. Which players' performances do you closely follow and engage with during matches?	0.523	
24. Have you noticed any changes in your engagement patterns with the team over multiple seasons?	0.659	Engagement pattern (time/season)
8. How do time zone differences impact your ability to engage with live matches?	−0.643	
23. Are there any cultural or local factors that influence how you engage with the team's content and events?	0.525	

(continued)

Table 5 (continued)

Variables	Factor loading	Factor label
1. Which city or region do you primarily follow the Chennai super kings from?	0.775	Location (city and region)
19. Do you find yourself more engaged during certain times of the year, such as playoffs or festive seasons?	0.517	
25. Are there any sports teams or leagues that you also engage with, and how does your engagement compare between them?	0.425	

4.3 *Relative Importance of the Factors Affecting fan Engagement*

The factor scores were correlated with the overall fan engagement score to drive the relative importance of the factors. The squared correlations were normalized so that the sum of weights adds up to 100.

As can be seen, the two key factors for fan engagement are (1) live interaction (58.5%) and (2) emotional connect (36.6%). The other two factors identified are (3) engagement pattern based on different times and seasons affecting fan involvement (3.0%) and (4) location (1.8%). These factors have also emerged as significant in the sentiment analysis carried out using social media comments from CSK fans.

Demographic analysis can help segment fans into groups for targeted engagement strategies. Younger fans might respond well to social media campaigns, while older fans might prefer personalized emails. Understanding demographic preferences can help tailor content to match the interests of different segments. For example, older fans might appreciate content that highlights team history, while younger fans might prefer content about current players and trends.

Hence, *live interactions*, both during matches and in off-match discussions, serve as crucial avenues for fostering a stronger connection with fans. This direct engagement allows for real-time interaction and feedback, contributing to a sense of unity and emotional attachment to the team. The concept of *team spirit and emotional connectivity* plays a pivotal role in fan engagement. When fans feel emotionally connected to the team, it creates a deeper bond and loyalty, driving them to support and engage with the team's activities actively. Studying the *patterns of engagement* over different times and seasons provides valuable insights into when and how fans are most active and interested. Understanding these patterns helps tailor strategies for consistent engagement throughout the year. Considering the *geographical aspect*, the city and region where the team is based and has a strong following can significantly impact engagement. Considering the specific demographics, preferences, and characteristics of the local fan base can guide targeted engagement efforts and strengthen the team's presence within the community (Table 6). Additionally, natural language processing is used to support the factors identified for fan engagement, and sentiment analysis is applied to the data scraped from social media.

Table 6 Importance of four factors

Correlations	r	r^2	Importance
Live interactions (online and offline)	0.754	0.568516	58.5
Emotional connect	0.596	0.355216	36.6
Engagement pattern (time/season)	0.172	0.029584	3.0
Location (city/region)	0.134	0.017956	1.8
Total		0.971272	100.0

4.4 Sentiment Analysis/Opinion Mining of the Fans

NLP is used to understand the emotional tone behind fans' statements. Data were collected from Facebook, the official websites of CSK, Twitter, CSK fan Instagram, and through common searches on Google (Sports Fan Engagement in IPL, IPL official website, Chennai Super Kings has highest fan loyalty: Rediffusion Red Lab) using Search Engine Optimization. We trained the algorithm to determine the polarity as to whether their sentiments are positive, neutral, or negative by analyzing the text they provide. This analysis provides insights into how fans feel about the team's engagement efforts. Positive ideas highlight strengths to build upon; negative sentiments identify areas for improvement, and neutral sentiments establish a baseline. This information guides decision-making to enhance fan engagement strategies effectively. We have taken a positive average and the count of the word is given in Table 7. The count of positive sentiments (428) is much higher than the negative sentiments (10) and the neutral count (32). The percentage occurrences of keywords that are frequently used by the fans are presented in Table 8.

The word count IPL, Dhoni, CSK, MS, Super, and Chennai have the highest percentage. So the fans are emotionally connected to these words. Team and Key players are the most significant words we can find in the word cloud shown in Fig. 4.

The data is clustered based on the textual content, and the word cluster results are shown in Fig. 5.

The words that pop up in the word cloud and cluster express the feelings of the people in a pictorial form. The picture that emerges from the word cloud and cluster tallies with the factors found using quantitative analysis. Additionally, the emotions captured are of real-time and the fans are seriously immersed in the team and its top players. When their team wins, the feeling is that the entire fans too have won. It's more than just cheering; it's like they're part of it.

The researchers observed through the visualization method that a significant portion of the fan base is distributed across several age buckets, notably within the range of 18–24, 25–34, 35–44, and even including those below 18 years old, as shown in Fig. 6. The percentage of fans in the age bucket of 20–30 is 49%, 25–34 is 35%, 35–44 is 12%, and under 18 is 2%. The different age groups might engage with sports teams differently. Younger fans might be more active on social media and engage with interactive content, while older fans might prefer traditional media or offline events. Age can influence opinions about the team's performance, players,

Table 7 Average polarity count for fan text

Average	Count	Polarity
0.98973281	428	Positive
−0.43552	10	Negative
0	32	Neutral

Table 8 Word count of the text of the fan in social media

	Count
IPL	48.0
CSK	47.0
Dhoni	25.0
MS	20.0
Super	18.0
Kings	17.0
Champion	14.0
Chennai	11.0
Champions	10.0
Jadeja	10.0
Sixes	8.0
Best	8.0
Gaikwad	8.0
One	8.0
Back	8.0
Msdhoni	7.0
Team	7.0
Year	7.0
Comeback	7.0
Ravindra	7.0

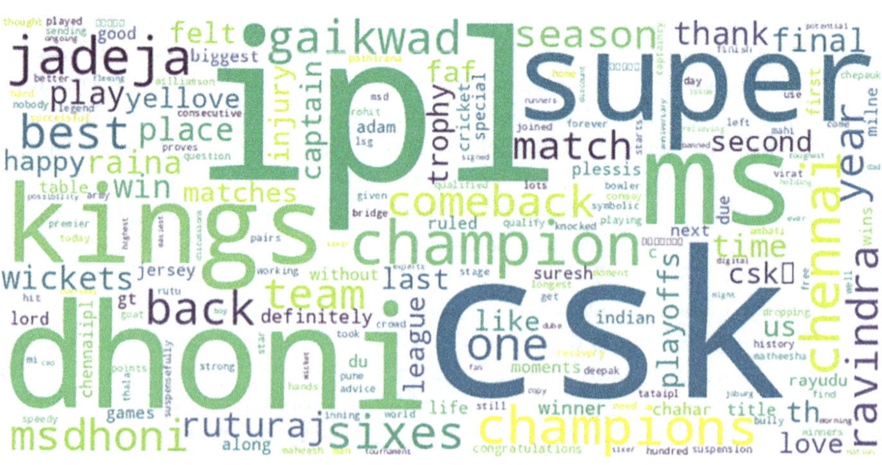

Fig. 4 Word cloud for the CSK fan text

Fig. 5 Word cluster

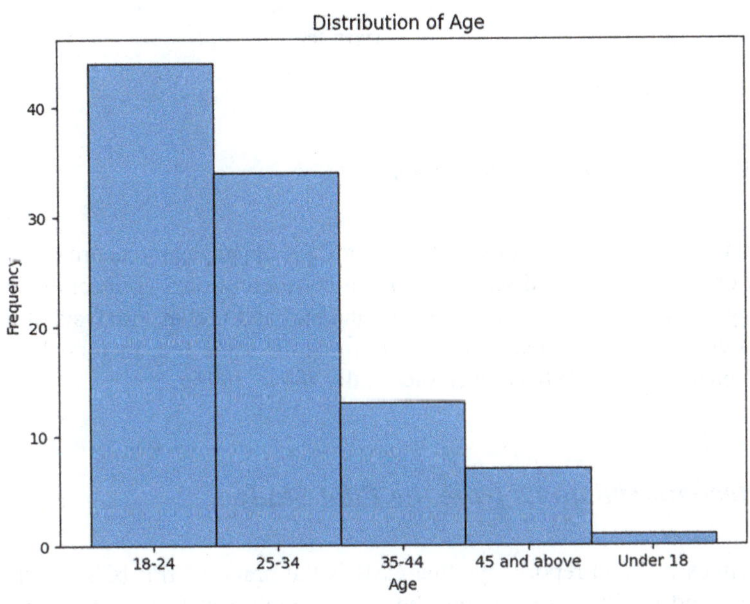

Fig. 6 Distribution of age in CSK fan engagement

and strategies. Younger fans might focus on newer players and trends, while older fans might have a deeper historical perspective.

Overall engagement level of the fans during a match is high, as shown in Fig. 7. The engagement level of more than 65% of the fans is either high or very high. Therefore, directing attention toward understanding and enhancing fan engagement levels could potentially facilitate CSK's journey toward reaching new heights across diverse segments.

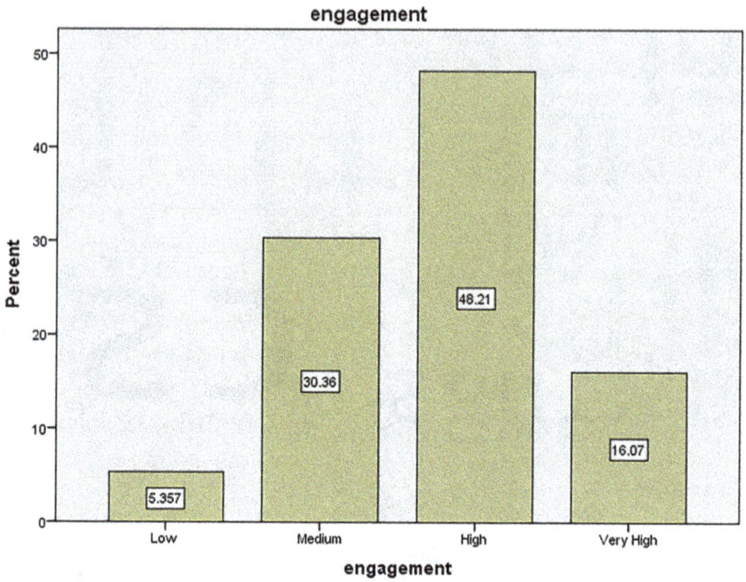

Fig. 7 Engagement analysis

4.5 Fan Engagement Model for CSK

The four factors that turned out significant for CSK fan engagement are live interactions in on and off match discussions, team spirit/emotional connection with the team, engagement patterns (time/session), and city and region. Analysis of qualitative data using sentiment analysis too confirmed the same four factors. The model of fan engagement for CSK is given pictorially shown in Fig. 8.

4.6 Recommendations from the Pilot Study

As the Times of India report says that CSK is the brand of IPL (CSK is the most valuable brand in IPL), considering these four significant factors, several recommendations could be made to CSK's brand promoters to enhance fan engagement and brand loyalty further:

- *Maximize Live Interactions:* Expanding live interactions during matches and off-match discussions should be a priority. This can include real-time Q&A sessions, live chats, and interactive social media campaigns. Creating an immersive fan experience during matches and fostering ongoing conversations between matches can forge a deeper emotional connection with the fan base.

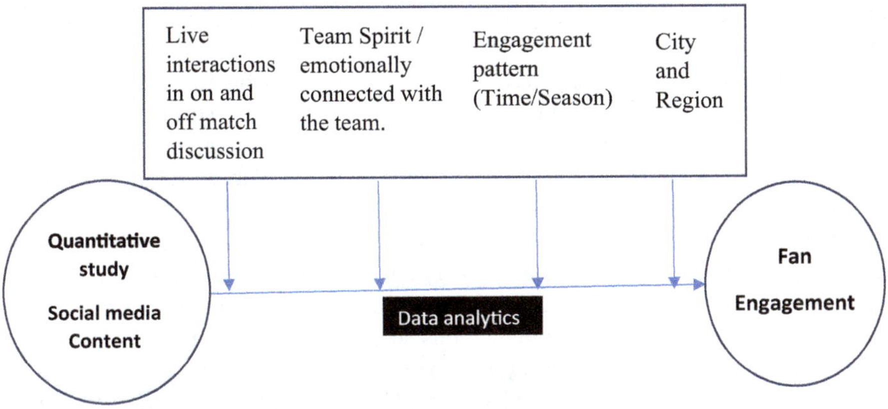

Fig. 8 Factors influencing fan engagement. (Data Source: Author)

- *Nurture Emotional Bond:* Capitalize on the team's spirit and emotional connectivity with fans. Building narratives around the team's journey, values, and achievements can intensify fans' emotional attachment. Heartfelt stories about players and behind-the-scenes content can help fans relate to the team more personally.
- *Tailor Engagement Strategies:* Analyzing engagement patterns over different times and seasons is key. Customizing content and engagement initiatives to align with peak engagement periods ensures consistent fan involvement. This could involve targeted campaigns during anticipated high-engagement times and providing ongoing interactions during off-seasons.
- *Localized Fan Engagement:* Recognizing the city and region's influence on engagement and fostering a strong local presence is crucial. Tailor content and initiatives that resonate with the specific characteristics and preferences of the local fan base. Community events, partnerships with local organizations, and region-specific campaigns can help solidify the team's connection with its hometown supporters. Incorporating these recommendations into the brand's promotional strategies can lead to heightened fan engagement, stronger brand loyalty, and a more profound connection between CSK and its devoted fan base.

5 Conclusion

This research presents a compelling exploration into the transformative role of data analytics within the dynamic realm of sports management. With a specific focus on the case of the Chennai Super Kings (CSK), the study adeptly navigates the landscape of fan engagement and satisfaction, shedding light on the potential that analytics holds for strategic optimization. The primary objective of enhancing fan experiences and driving revenue growth is magnificently addressed through an intricate methodology encompassing literature review, empirical data analysis, and qualitative insights. The study's findings resoundingly validate the profound impact that analytics methodologies can have on deciphering fan behavior and preferences, thus guiding the formulation of tailored engagement strategies. Through a deep dive into the CSK case study, this research illuminates the symbiotic relationship between analytics-driven insights and strategic decision-making. The correlation between effective fan engagement and subsequent revenue generation offers a pivotal revelation that can revolutionize sports management practices. The research is underscored by its originality, particularly in its laser-focused examination of IPL fan engagement, setting a new benchmark in the domain. Ultimately, the study is a beacon of guidance for sports organizations seeking to navigate the intricate maze of fan engagement. By underscoring the practical value of analytics and its transformative potential, this research establishes itself as a cornerstone reference for professionals aiming to optimize strategies and elevate fan experiences in the ever-evolving sports landscape.

References

Anderson, B., & Xu, Q. (2019). Winning over fans: how sports teams use live-tweeting to maximize engagement. *In The Golden Age of Data* (pp. 182–193). Routledge.

Anuraj, A., Boparai, G. S., Leung, C. K., Madill, E. W., Pandhi, D. A., Patel, A. D., & Vyas, R. K. (2023). Sports data mining for cricket match prediction. In *International Conference on Advanced Information Networking and Applications* (pp. 668–680). Springer International Publishing.

Bedeley, R. T., Benbow, M., & Iyer, L. S. (2016). *An analytic-based strategy for college sports ticket sales and management*. In 22nd Americas Conference on Information Systems (AMCIS) 2016 Proceedings. [ISBN: 978-0-9966831-2-8].

CSK Most Valuable Brand in IPL. (n.d.). *Sportskeeda*. https://www.sportskeeda.com/cricket/csk-most-valuable-brand-in-ipl

Glebova, E., Gerke, A., & Book, R. (2023). The transformational role of technology in sports events. In *Sports Management in an Uncertain Environment* (pp. 169–187). Springer Nature.

Gupta, S., & Divekar, S. (2022). A symmetry or asymmetry: Reflecting upon realities of cultural practice in sport psychology. *Sport and Exercise Psychology Review, 17*(1), 60–72.

Gupta, S., Bag, S., Modgil, S., de Sousa Jabbour, A. B. L., & Kumar, A. (2022). Examining the influence of big data analytics and additive manufacturing on supply chain risk control and resilience: An empirical study. *Computers & Industrial Engineering, 172*, 108629.

Hoeber, O., Hoeber, L., El Meseery, M., Odoh, K., & Gopi, R. (2016). Visual twitter analytics (vista) temporally changing sentiment and the discovery of emergent themes within sport event tweets. *Online Information Review, 40*(1), 25–41. https://doi.org/10.1108/OIR-02-2015-0067

Huettermann, M., Uhrich, S., & Koenigstorfer, J. (2022). Components and outcomes of fan engagement in team sports: The perspective of managers and fans. *Journal of Global Sport Management, 7*(4), 447–478.

Jakar, G., & Carr, J. (2023). A comparative analysis of social media fan (community) engagement in a European and a north American pro-sport league and their reaction to industry-wide disruptions. *American Behavioral Scientist, 67*(11), 1267–1288. https://doi.org/10.1177/00027642221118269

Jiang, K., Izadi, M., Liu, Z. Y., Dong, J. S., Li, Y., & Liew, A. (2020). *Deep learning application in broadcast tennis video annotation* (pp. 53–62). In 2020 25th International Conference on Engineering of Complex Computer Systems (ICECCS 2020). [ISBN: 978-1-7281-8558-3].

Kaur, A., Kaur, R., & Jagdev, G. (2021). Analyzing and exploring the impact of big data analytics in sports sector. *SN Computer Science, 2*(3), 184.

Kennedy, H., Kunkel, T., & Funk, D. C. (2021). Using predictive analytics to measure effectiveness of social media engagement: A digital measurement perspective. *Sport Marketing Quarterly, 30*(4), 265–277. https://doi.org/10.32731/SMQ.304.1221.02

Kim, J. O., & Mueller, C. W. (1978). *Introduction to factor analysis: What it is and how to do it* (Vol. 13).

Laurell, C., Liu, D. F., & Soderman, S. (2021). Tracking the international discourse of the Chinese super league through social media analytics. *Sport in Society, 38*(1), 6–17. https://doi.org/10.1080/17430437.2021.1916245

Li, B., Naraine, M. L., Zhao, L., & Li, C. (2023). A magic "Bullet": Exploring sport fan usage of on-screen, ephemeral posts during live stream sessions. *Communication & Sport, 11*(2), 334–355. https://doi.org/10.1177/21674795211038949

Mamo, Y., Su, Y., & Andrew, D. P. (2022). The transformative impact of big data applications in sport marketing: Current and future directions. *International Journal of Sports Marketing and Sponsorship, 23*(3), 594–611.

McCarthy, B. (2013). Consuming sports media, producing sports media: An analysis of two fan sports blogospheres. *International Review for the Sociology of Sport, 48*(4), 421–434. https://doi.org/10.1177/1012690212448240

Melnick, M. J., & Wann, D. L. (2011). An examination of sport fandom in Australia: Socialization, team identification, and fan behavior. *International Review for the Sociology of Sport, 46*(4), 456–470. https://doi.org/10.1177/1012690210380582

Morgulev, E., Azar, O. H., Lidor, R., Flanagan, et al. (2018). Sports analytics and the big-data era. *International Journal of Data Science and Analytics, 5*, 213–222.

Muniz, M., & Flamand, T. (2022). Sports analytics for balanced team-building decisions. *Journal of the Operational Research Society*, 1–18.

Naraine, M. L., & Bakhsh, J. T. (2022). Optimizing social media engagement in professional sport: A 3-year examination of Facebook, Instagram, and Twitter posts. *International Journal of Sport Communication, 15*(2), 103–116. https://doi.org/10.1123/ijsc.2021-0079

Noorbhai, H. (2022). Cricket coaching and batting in the 21st century through a 4IR lens: A narrative review. *BMJ Open Sport & Exercise Medicine, 8*(3), e001435.

Piché, M. C., & Naraine, M. L. (2022). Off the court: Examining social media activity and engagement in women's professional sport. *International Journal of Sport Communication, 15*(1), 23–32.

Santos, T. O., Correia, A., Biscaia, R., & Pegoraro, A. (2019). Examining fan engagement through social networking sites. *International Journal of Sports Marketing and Sponsorship, 20*(1), 163–183.

Scibilia, M., & Hutchins, B. (2011). High-stakes television: Fan engagement, market literacy and the Battle for sports content. *Media International Australia, 141*(1), 26–37. https://doi.org/10.1177/1329878X1114100105

Shuv-Ami, A., & Toder Alon, A. (2023). How do you categorize yourself as a sports fan? A new scale of sports Fan Social–Personal Identity Salience (FSPIS) and its consequences. *Communication & Sport, 11*(1), 28–52. https://doi.org/10.1177/2167479520967278

Sikka, R. S., Baer, M., Raja, A., Stuart, M., & Tompkins, M. (2019). Analytics in sports medicine: Implications and responsibilities that accompany the era of big data. *JBJS, 101*(3), 276–283.

Weimar, D., Soebbing, B. P., & Wicker, P. (2021). Dealing with statistical significance in big data: The social media value of game outcomes in professional football. *Journal of Sport Management, 35*(3), 266–277. https://doi.org/10.1123/jsm.2020-0275

Wikipedia Page: Chennai Super Kings in 2023. https://en.wikipedia.org/wiki/Chennai_Super_Kings_in_2023

Wu, C. W., Shieh, M. D., Lien, J. J. J., Yang, J. F., Chu, W. T., Huang, T. H., et al. (2022). Enhancing fan engagement in a 5G stadium with AI-based technologies and live streaming. *IEEE Systems Journal, 16*(4), 6590–6601. https://doi.org/10.1109/JSYST.2022.3169553

Yoshida, M., Gordon, B., Nakazawa, M., & Biscaia, R. (2014). Conceptualization and measurement of fan engagement: Empirical evidence from a professional sport context. *Journal of Sport Management, 28*(4), 399–417.

Sports Analytics in Operative Football Coaching

S. Binthu Mathavan and A. Praveen

1 Introduction

Sports analytics is a fast-expanding field that studies how to employ artificial intelligence (AI) and data analysis to enhance athletic performance. Understanding player and team performance and making data-driven decisions that can enhance game outcomes entails gathering, analyzing, and interpreting data. A wide range of sports, including football, basketball, baseball, and soccer, are using sports analytics more and more frequently. Teams can use this technology to their advantage by using it to better analyze their strengths and weaknesses, pinpoint areas for development, and make tactical choices that will increase their chances of winning. A wide range of sports, including football, basketball, baseball, and soccer, are using sports analytics more and more frequently. Sports data analytics is a crucial component of off-field soccer decision-making. To help players perform better on the field and make better decisions off the field, soccer teams all over the world have made significant investments in data science and associated technologies. Sports analytics can be used to monitor player progress, spot trends, and forecast results in the future. In addition, it can be used to evaluate the performance of the team, pinpoint areas in need of development, and make calculated choices regarding player choices and game strategies.

The act of gathering, evaluating, and interpreting data to obtain knowledge and make wise judgments in the sports industry is known as sports analytics. Numerous sources, such as player statistics, game footage, and even wearable technology, may

S. Binthu Mathavan (✉)
Department of Physical Education and Sports, Central University of Tamil Nadu, Tiruvarur, TN, India
e-mail: binthumathavan@cutn.ac.in

A. Praveen
Department of Physical Education and Sports, Pondicherry University, Puducherry, India

A. Mansurali et al. (eds.), *Sports Analytics*,
https://doi.org/10.1007/978-3-031-63573-1_7

provide this data. Teams can use sophisticated statistical models and algorithms to find patterns and trends that can guide their strategic decisions on the court or in the game. In sports, every team is constantly trying to find ways to improve and advance to the next level of competition. A vital tool for teams looking to boost performance and make wise judgments in this age of ever-evolving technology and data analysis is sports analytics.

Davenport and Harris (2007) classify analytics as descriptive, predictive, or prescriptive. Davenport and Kim (2013) highlight three stages of analytical thinking: problem framing, problem-solving, and communication and action on results. Professional sports clubs are increasingly utilizing analytical tools in ticket-selling strategies, sponsorship returns, and relationship management. As technology alters the knowledge used to make sporting decisions, analysis is becoming increasingly important for businesses operating in today's highly competitive corporate world (RM Silva, 2016). Whatever the reason, "sports analytics" is more often used than "sports statistics," possibly because its expertise comes from other fields such as Computer Science, Management Statistics, and Health Sciences. The common definition of sports analysis thus is the process of managing data, using forecast models, and making decisions that take advantage of information technology to gain a competitive edge.

In 1952 Mosteller conducted a study on the subject of evaluating the likelihood that the superior club will win the World Series in Major League Baseball. To promote the growth of statistics and their application in sports, an independent section on "Sports Statistics" was established by the American Statistical Association (ASA) in 1992. *Abstract Baseball*, a yearly baseball book written by Bill James, was originally published in 1977 and is widely regarded as the birth of sports analysis. Even though the majority of his literature did not employ the most basic statistical procedures, such as model fitting and graphical presentations, his work had a considerable impact on challenging long-held baseball beliefs. His data-supported opinions have been the distinguishing feature of his work. *Moneyball*, a 2003 book by Michael Lewis, explains how the Oakland A's, a small-market MLB franchise, managed to compete without high payroll by using various statistical methodologies. Thc 2011 film "Moneyball" adapted the book, enhancing appreciation for Lewis' work.

Sports Analytics (AI) in Soccer Games In the unbridling accuracy and ever-evolving empire of football coaching, the dynamism of brilliance and quality is a never-ending process. Since the game has become faster, more tactical and technical, and all-time challenging and competitive, the coaches are using all possible ways to find new adoptions and tools to understand the thoughtfulness of the game. At the vanguard of this revolution is the amalgamation of sports analytics into operative soccer coaching. Sports coaching is a scientific art that enhances performance in relevant sports and achievement in sports at an elite level; athletes require skill, technique, and tactical quality. In the last decade, video analytics has transformed playing patterns and training methods by utilizing technology to collect large, specific data sets related to sports activities cost-effectively and meticulously reviewed with experts to improve individual and team performance, especially in soccer games.

This chapter probes into the absorption of joining technology data and on-the-field and off-the-field approaches and tactics, reconnoitering how sports analytics is restructuring the landscape of football coaching. In the realm where every pass, every shot, every dribble, every attack, and every defensive maneuver can be examined and measured, football coaches now have access to a treasure trove of judgments that can improve their policy-making progressions. Sports coaching is a scientific art that enhances performance in relevant sports and achievement in sports at an elite level; athletes require skill, technique, and tactical quality. In the last decade, video analytics has transformed playing patterns and training methods by utilizing technology to collect large, specific data sets related to sports activities cost-effectively and meticulously reviewed with experts to improve individual and team performance, especially in soccer games.

This chapter details tactical analytic methods and applications nurturing coaches' development and creating top-notch soccer teams. This chapter helps in evaluating the strengths and weaknesses of soccer players and comprehends the fundamental scientific-analytical training techniques that can be used to improve playing quality in soccer games.

2 Sports Analytics Procedures

The sports analytics about players' movements, teams' movements, players' positions, strengths and weaknesses of own team, strengths and weaknesses of the opponent, as well as video recordings of 3D motion, have been covered in this title using the following procedures.

Collective Data Analysis: Analytical procedure to classify the acquired data into the desired categories.

Dispersion of Team: Scattering the analytics process according to the outlined parameters required for analysis.

Description of Problems: Problems and individual player performance should be assessed for the team, and a position-specific playing quality report should be analyzed.

Individual Analysis: It is essential to assess each player individually to identify their strengths and limitations, which may be used to improve their performance.

Exploration of Data: Under the analytics findings, training must be planned, and practical application must be implemented to reduce weaknesses and increase player and team performance.

Implementation of Analytics: As per the results of analytics, the training needs to be prepared, and practical application needs to be implemented to minimize the weaknesses and to build the strength of individual players and team performance.

Evaluation of Analytics: Collect and examine data to assess or react to training programs that have been put in place for a soccer team that is concerned, as well as for individual progress, to determine the effects of the implemented training.

This chapter covered the process for collecting sports analytics data, group data analysis, problem description analysis, detecting weaknesses, implementing alternative training to improve team and individual performance, and evaluation. Soccer players, coaches, and sports analysts who wish to strengthen their squad to compete in higher-level competitions may find this factual.

3 Sports Analytics in Operative Football Coaching Information

The data is gathered in many ways, particularly in football. It can be gathered through conversations, observations, measurements, audio, video, numerical methods, etc. There is data available everywhere, but big data analytics is currently the most well-liked and developing one. With analytics, it is possible to analyze many sports-related parameters and make high-quality, functional decisions in relevant domains based on demand. The solicitation of analytics in football coaching goes beyond sheer numerical scrutiny. It involves the removal of eloquent strategies and leanings from cosmic data sets, supporting coaches with a wide-ranging view of their instrument as well as that of their opponents. From player tracking to tactical analysis, this chapter circumnavigates the multilayered scopes of sports analytics and its pivotal role in optimizing player development and team success These days, it's a very sought-after area in the sports industry, particularly in the highly watched sport of soccer worldwide. Sports analytics are utilized not only to draw in crowds but also to improve coaching errors, player quality, spectator contentment, sales in the sports sector, and the development of improved strategies and approaches for improving performance (Goes, 2021).

As we explore the pages ahead, we will unravel how coaches are leveraging data-driven insights to refine training methodologies, fine-tune tactical strategies, and make informed decisions in the heat of the game. We will observe in what way the conjugal of old-fashioned coaching wisdom with pioneering analytics is not just a luxury but a necessity in today's soccer. The data-gathering technologies listed below can be utilized to obtain information from teams and players. The journey that surpasses the limitations of orthodox coaching, where the connubial of sports and science is boosting football into a new era. The game is no longer drawn out just to lodge yourself on the field it reveals in the structures, outsets, and analytical facsimiles that escort coaches toward unlocking their team's full potential. Welcome to the era of sports analytics in operative soccer coaching—where precision meets passion, and every data point tells a story on the path to victory head-mounted displays (HMD), immersive video 360 0 VR, cave automatic virtual environment (CAVE), and virtual reality (VR) technologies (Kittel et al., 2020). A significant portion of research has been conducted using data collected from the video tracking system using observation methods, local position measurement (LPM), global positioning system (GPS), optical tracking methods (Wang and Parameswaran 2004), and visual analytics (Moher et al., 2015; Frencken et al., 2010; Mara et al., 2017). People's interest in sports video analytics has grown due to the advancement of fast, specialized digital cameras and video processing (Rui et al. 2000). It facilitates decision-making for referees in all sports by tracking individual players, balls, and team performance. Currently, one of the most sought-after methods in sports analytics is video monitoring (Xu et al., 2003). Several cameras must be used to monitor an item or collect data along a distinct trajectory to apply these techniques, which may provide a clear perspective for obtaining high-quality data for big data analytics (Yu et al., 2003). When compared to other forms of analytics, video tracking (Groom et al., 2011), which uses video monitoring technology to evaluate the behavior of soccer coaches and players both on and off the field, is more successful since it is evidence-based.

4 Collective Data Analysis

Football is a fast-paced sport, and success frequently comes from planned execution, painstaking preparation, and the capacity to quickly adjust to changing circumstances on the pitch. The use of data in football coaching has become increasingly important as the game enters a more advanced era. This chapter delves into the crucial element of collective data analysis and its revolutionary influence on practical football instruction. To get a complete picture of a football team's dynamics, collective data analysis entails a thorough analysis of player statistics, team-wide performance indicators, and tactical insights. As we move through this chapter, we will recognize that success on the pitch is frequently a joint endeavor, and it essentially signifies a shift away from individual player analysis and toward a more connected and collective approach.

As we proceed through this chapter, we will see how coaches use the strength of team data to craft methods that play to everyone's talents. Every facet of football, including player interplay and team formations, is examined through the prism of analytics to improve team performance as a whole. Collective data analysis shines a light on the field of operative football coaching, helping coaches make decisions. Combining team-wide statistics with individual player performance data offers a more complex picture than what is initially apparent. It enables coaches to spot trends, enhance playing methods, and create plans that play to the team's strengths as a whole.

The upcoming chapters will explore real-world scenarios and case studies that illustrate how football coaches use collective data to solve problems like cohesive teamwork, strategic synchronization, and game dynamics adaptation. The utilization of collective data becomes a dynamic instrument that coaches utilize to negotiate the complexity of modern football, from pre-match preparations to in-game changes.

In this, data analysis is collected and its format may vary depending on the needs of a person, group, team, club, association, or organization. This is particularly true in soccer coaching for skill performance analysis; the data will be numerical/measured in time, distance, force, etc. In this case, to analyze each athlete's physical and physiological qualities or as a team, experts may collect quantitative responses regarding psychological reactions, formerly considered qualitative (Filetti et al., 2017).

The following details about soccer teams' collective behavior may be discovered by studying collective data: speed, strength, endurance, coordination, agility, reactivity, flexibility, playing structure, playing strategy, playing quality, team mobility, tactical behavior, and competition-related injuries. Skills measured specifically passing, kicking, shooting, dribbling, free kicks, penalty kicks, heading corner kicks, etc. (Cotta et al., 2016; Silva and Marcelino 2023). With this knowledge, professionals must classify the collected data into many categories according to the demands of the study questions, such as psychological factors such as anxiety, aggression, arousal, self-esteem, self-confidence, task orientation, ego, performance satisfaction, and creative thinking. For example, before beginning any training camp, players' physical, physiological, and psychological data will be collected. The same data will be collected after 8 weeks or 3 months. With this information, we can assess the players' playing quality by comparing their pre- and post-experiment positioning for the ball, team cohesion, skill implementation and success rate, technique, and tactical implementation.

Most researchers use video scouting to gather data on soccer players; we can find various required information, and it must be trustworthy, consistent, valid, explicit, accessible, traceable, comprehensible, key performance indicator (KPI), and transparent (Fischer et al., 2019; Herold et al., 2021). Some processes are as follows: *Event data*: Only particular information about the event, such as the number of yellow cards, fouls, free throws, shooting attempts, door shots, and player performance in various competition situations, will be filtered here. *Follow-up Data*: A player's inside-the-field movement, speed, position, and tactical response are tracked as additional data. *Data on perception*: Specifics about each player used to be examined here, such as their passing accuracy and skill. *Context data:* In this environment, fundamental information about the weather, the date and location of the competition players' equipment, and how well it performed in various scenarios is used to examine. Large-scale data analysis presents many challenges in video scouting, particularly when analyzing soccer game data.

Mode and Category of Data The use of data analysis has become essential in the rapidly changing field of football coaching to maximize player performance, refine tactics, and extract insightful information. This chapter takes the reader on a tour through the complexities of data analysis, emphasizing the important components of modes and categories that influence how operational football coaches make decisions.

Football data analysis is a complex process that involves many modes that capture different aspects of the game. To uncover the subtleties that characterize success on the pitch, coaches are traversing a rich tapestry of data modes, from the microscopic examination of individual player motions to the macroscopic assessment of team-wide trends.

Data analysis methods go beyond classical statistics and include physiological indicators, spatial data, and video analysis. With the help of each mode's layer of information, coaches can create a thorough narrative about a team's performance. This chapter will explore how these modes interact, providing a synergistic method for comprehending and improving all facets of football coaching.

The classification of data is closely related to modes and contributes an additional level of complexity to the analysis. Coaches may make meaningful comparisons, spot trends, and customize methods to capitalize on particular strengths or correct deficiencies by grouping data into distinct categories. The classification of data, whether it be through player performance categories, tactical classifications, or situational assessments, acts as a guide for effective football coaching.

As we move through this chapter, case studies and real-world examples will help us understand how various data modes and categories are used in football coaching. The tactical application of data analysis is changing the face of operational football coaching, from pregame planning to postgame assessments. It is shown to be more than simply a tool, but a dynamic ally in the quest for victory.

Greetings from the complex realm where data analysis modes and categories join together to unravel the football playbook. Come along as we disentangle the nuances, investigate the connections, and learn how coaches traverse this data-driven landscape to craft tactics that go beyond the bounds of traditional coaching. Success in this field is determined not only by goals but also by perceptive analysis of the data that drives the football team's victories.

When classifying the data, consideration should be given to age, gender, level of competition (e.g., school, college, district, state, national, World Championship, Olympic Tournament, Premier League, AFC Challenge Cup, Afro-Asian Cup, ISL, official club matches, etc.) as these factors will determine the team's performance potential based on the tactical realities.

5 Dispersion of Team

Recognizing Dispersion in Team Analysis for Operational Football Coaching: Going Beyond the Huddle.

Football is a complex dance, and a team's ability to execute depends largely on how well its players coordinate their moves and situate themselves on the pitch. As operative football coaching explores more data-rich areas, team dispersion becomes apparent as a crucial factor that can inform tactics, maximize output, and impact results on game day. This chapter walks through the idea of team dispersion and examines how coaches can improve their tactical repertoire by using this data-driven indicator.

The spatial organization and distribution of players on the pitch at any particular time is referred to as team dispersion. The dispersion of a team is more than just a pretty picture; it's a dynamic statistic that takes into account player positioning, offensive and defensive strategy, and the team's ability to take advantage of or counteract the game's spatial dynamics.

This chapter will explain the value of team dispersion in operative football coaching and show coaches how to use this statistic to get a clear picture of their team's tactical effectiveness. The dispersion of players becomes a canvas upon which coaches paint their strategic masterpieces, modifying their formations and tactics based on real-time spatial insights, from defensive solidity to offensive fluidity.

Coaches may quantify and visualize team dispersion through the lens of data analysis, exposing trends that could otherwise be missed in the heat of competition. Dispersion metrics research goes beyond positional analysis to provide a more sophisticated understanding of player movements, spacing, and the overall spatial organization of the team.

In the first phase, video scattering analytics estimation results will be used to assess the player's quality based on various skill-related abilities and team formation. This will be done by the demands of different positions and the individual qualities that are demonstrated in performance, as well as the evaluation of the soccer game expert who will be utilized to implement team dispersion (Low et al., 2020). The distribution will be determined by several significant findings, such as enhanced tactical understanding, the significance of playing position, the perception of teammates and rivals, player strength, synchronization and consistency in performance, and so forth.

Using a trial-and-error method may be necessary to identify a suitable player in the relevant position who can function as a key player at this time. This stage may yield discoveries about a range of sports-related functional domains, business interests, leadership qualities, social media analytics, game outcomes, social media marketing, spectator-based marketing, operational leadership, and more.

Machine learning-based data analytics in sports marketing are widely accepted in society (Nitin Singh, 2020); nevertheless, other factors such as team possession style, open play, and possession of the score box during counterattacks can also be taken into account (Glazier, 2010). A SWOT analysis is a tool used to assess the strengths, weaknesses, opportunities, and threats of a sports organization. It helps identify priorities, highlight opportunities, and manage vulnerabilities. It provides a comprehensive view of the organization's internal and external environments, enabling a comprehensive scenario analysis (Lllagjevic-Govori et al., 2019). In soccer game strength and weakness of both our team and of our opponents in terms of player possession, playmaking ability, defensive maneuvers, goalkeeping, corner kicks, penalty shots, fouls committed, number of yellow and red cards issued, free kicks, ball passing, and skills need to be scatter based on parameter. It is necessary to assess playing strategy and tactical play based on the previously stated talent to classify the players and team quality at different levels. We can execute an effective playing move in a competition once we evaluate these factors for both our team and our opponent based on the analytics data with standard parameters. The strengths, weaknesses, and threats are divided into two sections; a positive threat against the opponent's team and a negative threat of the own team are assessed based on the parameter.

6 Description of Problems

Problems are identified in various ways depending on the requirement of a parameter, such as the physical quality of players, physiological qualities, psychological qualities, technical and tactical qualities, team composition, competition outcome, and match strategy, scattering from big data (Baena, 2016). Statistical analysis, video analysis, stick figure analysis, and animated analysis should all be employed in this case to identify the issue. For comparison between the team and individual players and to determine the impact of a match or training session, the statistical tool Pearson's coefficient can be employed. Additionally, an independent "t" test can be used to assess the normality of data. Other tools, such as Shapiro and homoscedasticity, can be utilized (Filetti et al., 2017). According to Glazier (2010), an innovative method of performance analysis may be appropriate. The SWOT analysis was created by Albert Humphrey and replaces threats with challenges. To help B. T. Solov evaluates conditions and makes managerial decisions in the field of sports, researchers wish to apply SWOT or SWOC analysis for strategic management (Dyakova and Dyakov 2019). According to earlier research findings (Shaker et al., n.d.; Prieto et al., 2021; Harridon et al., 2022), it will be more acceptable to group the issues based on the analytics into four branches: strengths, weaknesses, opportunities, and challenges (SWOC/SWOT) of an individual player or team.

7 Individual Analysis

In football coaching, individual analysis goes beyond simple statistical analysis to include a detailed look at a player's actions, choices, and individual contributions to the team dynamic. This chapter sets out to explore the layers that constitute modern game quality by delving into the wide world of player-centric data.

Coaches use individual analysis to create training plans that are tailored to each player's unique needs, from assessing technical proficiency to comprehending the cognitive components of play. When it comes to evaluating offensive flair, defensive contributions, or throwing accuracy, coaches can use the detailed information gathered from individual player analysis as a compass to help them improve their players and build successful teams.

We will examine the mutually beneficial relationship between team dynamics and individual performance in this chapter. Case studies and real-world examples will show how coaches combine player-centric statistics into a more comprehensive plan, striking a balance between the science of effective team play and the art of developing individual talent.

The storyline of football is shaped by each player as the game progresses. Greetings from a world where every move, step, and choice is not just observed but also carefully examined. Come along as we explore the nuances of individual analysis in practical football coaching, where player performances merge to create the winning rhythm on a big stage.

It's one of the most crucial parts of the game where each player is assessed individually based on a variety of factors, including their level of physical fitness, the quality of their soccer-related skills (such as kicking, passing, dribbling, and positioning the ball), the team's combination, the positioning of its players, and their movement, as well as their strategies for dealing with various situations and their tactical approach, as well as their strengths and weaknesses in various situations. SWOC analysis must be used to filter the individual performance analytics based on strengths, weaknesses, opportunities, and challenges (SWOC); by doing so, it will be simple to measure individual growth and select quality players for a stronger team in a competition (Oh et al., 2021; Dyakova and Dyakov 2019). This could make it easier to build an efficient training schedule where the person's quality can be stimulated.

8 Exploration of Data

Compile all available match data, such as possession percentage, shots on target, shots off target, completed passes, etc.:

(a) Analyze data according to player, position, and team to pinpoint advantages and disadvantages.
(b) Player tracking: Track player movements during practices and games using GPS and other tracking devices.
(c) To evaluate playing style and fitness, analyze sprint distance, speed, accelerations, and decelerations.
(d) Video analysis: To evaluate individual and team performances, record and examine match footage. Utilize tools for video analysis to dissect player actions, tactical choices, and important moments.
(e) Injury data: Monitor injuries and recuperation periods to modify exercise schedules and stop recurrent problems.

9 Interpretation of Data

(a) Analytical Strategy: Dissect the playing styles, tactics, and formations of the team. Examine opposition strategies to find openings and dangers.
(b) Metrics for Player Performance: Analyze each player's performance individually using statistical metrics. Analyze the key performance indicators (KPIs) that are pertinent to each position, such as successful dribbles for forwards and interceptions for defenders.
(c) Analysis of Set Pieces: Analyze the efficacy of set-piece tactics in both offensive and defensive scenarios. Determine trends and areas that could want improvement.
(d) Analysis of Time and Space: Examine how players use space and time in the course of a match.
(e) Determine the regions of the field where the team performs well or poorly.

10 Application of Data

(a) Training Program Modification: Adapt training regimens in light of fitness and performance metrics.
    Create drills that target particular areas of weakness or build on strengths.
(b) Opposition Analysis: Using data, create tactics to take advantage of future opponents' vulnerabilities.
    Create set-piece routines by analyzing the weaknesses of the opposition.
(c) Player Development: Determine each player's areas for growth and design tailored development strategies.
    Give helpful criticism based on performance information.
(d) Making Decisions During the Game: During a match, use real-time data to modify tactics.
    Track player weariness levels to schedule substitutions for the best possible times.

Operational football coaching analysis is a continuous process that includes data collection, interpretation, and application to improve team performance in several areas. This methodology's effectiveness is attributed to its integration of technology, ongoing feedback, and a comprehensive approach to player and team analysis.

The collective behaviors of soccer players need to be categorized by the results of big data analysis, and appropriate actions need to be taken to fix mistakes and improve performance in various skills, techniques, and tactical abilities (Aquino et al., 2016; Duarte et al., 2012). Based on the information gathered, an experimental task needs to be scheduled. In addition, the teams' physical condition must be appropriate for the setting and weather of the competitions, and training must be planned accordingly. Before beginning the scheduled training, it is possible to finalize the percentage of training intensity for various demands as well as taking into account opponents' strengths and weaknesses to overcome or challenge. To test the effectiveness of the given training, we must ensure the quality of the experiment's implementation and evaluation. To do this, we can use accepted scientific testing methods to examine each skill quality that is necessary for a soccer player (Duarte et al., 2012). The strengths, weaknesses, opportunities, and challenges (SWOC) of a player as well as a team need to be examined in the obtained data, and this is where the majority of the research has been done (Oh et al., 2021; Pan et al., 2017; Dyakova and Dyakov 2019; Li et al., 2020). It is important to categorize each player's strengths according to their playing positions and the necessary skills, techniques, and tactical demands so that professionals can clearly understand where, what, when, why, and how they must work. Only then can effective training be designed to produce high-quality players and teams to produce quality results.

11 Implementation of Analytics

To apply analytics to operative football coaching, data-driven insights must be integrated into all facets of coaching, including as player development, in-game decision-making, match preparation, and overall team strategy. This is a step-by-step explanation explaining how analytics can be used in football coaching:

(1) Establish Goals: Clearly state the aims and purposes of incorporating analytics into football coaching. Determine which particular domains—tactical analysis, player performance assessment, injury prevention, and opponent scouting, for example—are most impacted by analytics.
(2) Data Collection and Integration: Create a strong system for gathering information on matches, players, videos, and other pertinent metrics. Combine information from several sources to provide a complete picture of the performance of the team and the players.
(3) Assemble a Team for Data Analysis: Put together a group of data analysts with experience in sports science, statistics, and football analytics. Make sure the team works closely with the coaching staff to comprehend the squad's unique wants and specifications.
(4) Make Use of Advanced Metrics: Go beyond simple statistics and use advanced metrics to gain a deeper understanding of individual and group performance. Take into account indicators such as player heat maps, expected goals (xG), pass completion rates in particular areas, and pressing intensity.
(5) Video Analysis: Invest in tools that help coaches analyze match videos and pinpoint crucial moments. Review player actions, tactical choices, and opponent tactics using video analysis.

The results of the analytics must be taken into consideration while designing training programs and practical applications to minimize weaknesses and optimize strengths for both individual players and the team's performance. Coaches must balance faithfully reproducing match-play circumstances with ensuring that students of all ages and abilities receive the right attention demands (Ford et al., 2010). In soccer game, SWOC/SWOT analytics is most popular in current days where most elite soccer teams implement that the study uses an enterprise architecture methodology to assess Manchester United Football Club's present performance in (Azhar et al. 2022). Following the data analytics, we have a range of opinions and ideas on what sort of requirements we have in terms of the team, club, association, and players the decision should be made based on. Analytical implementation is an action plan based on data and demands that we intend to improve; these demands might be related to players' tactics, techniques, and management abilities as well as sports business, competition setup, team organization, and internal team dynamics (Nasution et al., 2023).

12 Conclusion

The chapter has aimed to examine coaching analytics as they currently exist and to clarify the processes involved in putting together a competitive football team in this chapter. This chapter may be of interest to football fans, managers, players, coaches, and other sports marketers. Researchers looked at the SWOC/SWOT technique in this study because it is highly effective compared to other approaches and is currently in high demand in all industries, not only sports.

The chapter talks about how analytics can assist football players and teams in becoming better players. By employing SWOC/SWOT techniques and high-quality training experiments, coaches and players can create elite teams with the help of sports analytics (AI). The many big data analytics methods, such as data scattering and video scouting, help examine a variety of factors.

> *Success isn't a coincidence. It requires a lot of effort, tenacity, sacrifice, and, most importantly, passion for what you are doing.*—Pelé

References

Aquino, R. L., Gonçalves, L. G. C., Vieira, L. H. P., Oliveira, L. P., Alves, G. F., Santiago, P. R. P., & Puggina, E. F. (2016). Periodization training focused on technical-tactical ability in young soccer players positively affects biochemical markers and game performance. *The Journal of Strength & Conditioning Research, 30*(10), 2723–2732.

Azhar, A. A. A., Hisham, B. E. B., & Bakar, N. A. A. (2022). Using the enterprise architecture approach to analyse the current performance of Manchester United Football Club. *Journal of Techno-Social, 14*(1), 28–36.

Baena, V. (2016). Online and mobile marketing strategies as drivers of brand love in sports teams: Findings from Real Madrid. *International Journal of Sports Marketing & Sponsorship, 17*, 202–218.

Cotta, L., de Melo, P. O. V., Benevenuto, F., & Loureiro, A. A. (2016). *Using fifa soccer video game data for soccer analytics*. In Workshop on Large Scale Sports Analytics.

Davenport, T., & Harris, J. (2007). *Competing on analytics*. Harvard Business School Press.
Davenport, T., & Kim, J. (2013). *Keeping up with the quants*. Harvard Business School Press.
Duarte, R., Araújo, D., Freire, L., Folgado, H., Fernandes, O., & Davids, K. (2012). Intra-and inter-group coordination patterns reveal collective behaviors of football players near the scoring zone. *Human Movement Science, 31*(6), 1639–1651.
Dyakova, G., & Dyakov, T. (2019). Swot-analysis of organization for including a football game in The Educational Programme on Physical Culture and Sport for Female Students. *Trakia Journal of Sciences, 17*(1), 652–656.
Filetti, C., Ruscello, B., D'Ottavio, S., & Fanelli, V. (2017). A study of relationships among technical, tactical, physical parameters and final outcomes in elite soccer matches as analyzed by a semiautomatic video tracking system. *Perceptual and Motor Skills, 124*(3), 601–620. https://doi.org/10.1177/0031512517692904
Fischer, M. T., Keim, D. A., & Stein, M. (2019, October). Video-based analysis of soccer matches. In *Proceedings Proceedings of the 2nd International Workshop on Multimedia Content Analysis in Sports* (pp. 1–9).
Ford, P. R., Yates, I., & Williams, A. M. (2010). An analysis of practice activities and instructional behaviours used by youth soccer coaches during practice: Exploring the link between science and application. *Journal of Sports Sciences, 28*(5), 483–495.
Frencken, W. G., Lemmink, K. A., & Delleman, N. J. (2010). Soccer-specific accuracy and validity of the local position measurement (LPM) system. *Journal of Science and Medicine in Sport, 13*(6), 641–645.
Glazier, P. S. (2010). Game, set and match? Substantive issues and future directions in performance analysis. *Journal of Sports Medicine, 40*(8), 626–634.
Goes, F. R., Meerhoff, L. A., Bueno, M. J. O., Rodrigues, D. M., Moura, F. A., Brink, M. S., & Lemmink, K. A. P. M. (2021). Unlocking the potential of big data to support tactical performance analysis in professional soccer: A systematic review. *European Journal of Sport Science, 21*(4), 481–496.
Groom, R., Cushion, C., & Nelson, L. (2011). The delivery of video-based performance analysis by England youth soccer coaches: Towards a grounded theory. *Journal of Applied Sport Psychology, 23*(1), 16–32.
Harridon, M., Abdullah, M. S. A. M., & Shah, S. A. (2022). Creation of passing lanes in soccer: On the surface observation based upon threats and opportunities of team and opposition. *International Journal of Scientific and Research Publications, 12*(2).
Herold, M., Kempe, M., Bauer, P., & Meyer, T. (2021). Attacking key performance indicators in soccer: Current practice and perceptions from the elite to youth academy level. *Journal of Sports Science & Medicine, 20*(1), 158. https://doi.org/10.52082/jssm.2021.158
https://www.futbollab.com/en/news/optimize-your-scouting-with-football-video-analysis
Kittel, A., Larkin, P., Cunningham, I., & Spittle, M. (2020). 360 virtual reality: A SWOT analysis in comparison to virtual reality. *Frontiers in Psychology, 11*, 563474.
Li, H., Chen, X., Chen, S., & Yang, K. (2020). Research on sustainable development of Chinese campus football based on SWOT analysis. *Insight-Sports Science, 2*(2).
Lllagjevic-Govori, A., Tahiraj, E., Gashi, A., & Llagjaj, D. (2019). Swot analyse of football school "2 korriku". *Sport Mont, 17*(1), 87–90.
Low, B., Coutinho, D., Gonçalves, B., Rein, R., Memmert, D., & Sampaio, J. (2020). A systematic review of collective tactical behaviours in football using positional data. *Sports Medicine, 50*, 343–385.
Mara, J., Morgan, S., Pumpa, K., & Thompson, K. (2017). The accuracy and reliability of a new optical player tracking system for measuring displacement of soccer players. *International Journal of Computer Science in Sport, 16*, 175–184.
Moher, D., Shamseer, L., Clarke, M., Ghersi, D., Liberati, A., Petticrew, M., et al. (2015). Preferred reporting items for systematic review and meta-analysis protocols (PRISMA-P) 2015 statement. *Systematic Reviews, 4*(1), 1–9.

Nasution, H., Rahayu, A., Wibowo, L. A., Dirgantari, P. D., Yulianto, E., & Nurgraha, R. (2023, September). Business strategy through SWOT analysis in implementing loan origination system to improve Bank's business performance (case study at mortgage credit). In *7th Global Conference on Business, Management, and Entrepreneurship (GCBME 2022)* (pp. 1593–1602). Atlantis Press.

Oh, Y. S., Arthur-Banning, S. G., & Domka, M. (2021). SWOT analysis on the potential growth of football 5-a-side programme across the United States: An exploratory case study approach for athletes with visual impairment. *Sport in Society, 24*(9), 1683–1697.

Pan, Y., Xu, C., & Yang, H. (2017, January). Research on application of SWOT analysis model in Chinese professional league. In *2016 2nd International Conference on Economics, Management Engineering and Education Technology (ICEMEET 2016)* (pp. 720–724). Atlantis Press.

Prieto, M. F., Gutiérrez, I. M., & Barrero, A. M. (2021). *Performance analysis of the tactical systems of a professional soccer* (pp. 111–120). E-Balonmano.com: Revista de Ciencias del Deporte.

Rui, Y., Gupta, A., &Acero, A. (2000, October). Automatically extracting highlights for TV baseball programs. In *Proceedings of the Eighth ACM International Conference on Multimedia* (pp. 105–115).

Shaker, R., Moharramzadeh, M., & Imanzadeh, M. Evaluation and analysis of East Azarbaijan football strategic plans and development of appropriate strategies using SWOT matrix. *Evaluation and Analysis of East Azarbaijan Football Strategic Plans and Development of Appropriate Strategies Using SWOT Matrix.*

Silva, R. M. (2016). *Sports analytics* (pp. 2–3). Doctor of Philosophy Thesis, Simon Fraser University.

Silva, H., & Marcelino, R. (2023). Inter-operator reliability of InStat scout in female football games. *Science & Sports, 38*(1), 42–46.

Singh, N. (2020). Sport analytics: A review. *The International Technology Management Review, 9*(1), 64–69. https://doi.org/10.2991/itmr.k.200831.001. ISSN 2213-7149; eISSN 1835-5269. https://www.atlantis-press.com/journals/itmr

Wang, J. R., & Parameswaran, N. (2004, June). Survey of sports video analysis: Research issues and applications. In *Proceedings of the Pan-Sydney area workshop on visual information processing* (pp. 87–90).

Xu, R. Y. D., Allen, J., & Jin, J. S. (2003). *Robust real-time tracking of non-rigid objects.* Conferences in Research and Practice in Information Technology, VIP'03.

Yu, X., Xu, C. S., Leong, H. W., Tian, Q., Tang, Q., & Wan, K. W. (2003). Trajectory-based ball detection and tracking with applications to semantic analysis of broadcast soccer video. In *Proc. of ACM MM'03* (pp. 11–20).

Unveiling the Dynamics: Kinematics and Activation Patterns of Gluteus Maximus in Inclined Treadmill Sprinting

K. Akshay, Babu Subbiah, and J. Ashokkumar

1 Introduction

Injuries such as strains, cramps, and muscle pull to the gluteus muscles and hamstring group of muscles are common in dynamic sports like football, basketball, field events, and sprinting. Several studies have been conducted on horizontal running and lower extremity muscles, running on different surfaces, at different paces. But there is still a knowledge gap in the impact of inclined running loading on muscle activation in adults. Some studies have been conducted, but more in-depth research is needed in this area because no one will be focused on strengthening the gluteus maximus (GM) which helps in the hip extension phase of the running gate.

The hypothesis proposed in this chapter with a pilot study which suggests that on inclined running with increasing repetitions, fatigue may result in decreased muscle activation, particularly in the gluteal and hamstring muscles. Recognizing the mani fold benefits of regular physical activity, including improvements in respiratory health, flexibility, and body composition, underscores the significance of strengthening lower extremity muscles, particularly the hip extensors, in light of the prevalent issue of gluteus strain injuries (Fiore, 2017; Manickavelu et al., 2022; Manickavelu & Kaliyaperumal, 2022). This underscores the significance of enhancing gluteal muscle activation for reinforcing the posterior chain. The investigation aims to assess the effect of a 12-min inclined running session on GM muscle activation. Gluteus strain injuries exhibit similar prevalence rates across various sports,

K. Akshay · B. Subbiah (✉)
Department of Physical Education, National Institute of Technology, Thiruvettakudy, Puducherry, India

J. Ashokkumar (✉)
Department of Management, School of Commerce and Business Management, Central University, Thiruvarur, Tamil Nadu, India
e-mail: ashokkumarj@cutn.ac.in

A. Mansurali et al. (eds.), *Sports Analytics*,
https://doi.org/10.1007/978-3-031-63573-1_8

with recent studies reporting 8–10% occurrence in Indian professional football, and 12% in soccer, contributing to hamstring-related injuries (Franettovich Smith et al., 2017).

Gluteus muscle strain injuries and muscle cramps are common challenges due to persistent symptoms, slow healing, high re-injury rates, uneven surface, and wrong practice especially when athletes lack eccentric strength training (Hawkins et al., 2001; Buckthorpe et al., 2019; Jeong et al., n.d.). Athletes experiencing their initial injuries, and stability due to lack of muscle strength specifically involving the anterior cruciate ligament (Kaiyaperumal et al., 2022, 2023) and gluteus, may encounter re-injury rates as elevated as 20% throughout an entire season by rehabilitation (Eliot et al., n.d.; Vasudevan et al., 2012). Various studies have delved into the mechanisms of gluteus muscle injuries, particularly during training overload and sprinting and short sprint training, aiming to enhance rehabilitation and injury prevention outcomes. Kara et al. (2015) discovered that individuals who suffer from hip pain, stiffness, and uneasy dynamic movements exhibited heightened amplitude in all gluteal muscle divisions, underscoring the significance of comprehending these muscle activation patterns. The significance of adopting a comprehensive strategy is acknowledged, with the integration of medication, meditation, and yoga in the various forms of physical pain in all the body muscles and joints (Jagadevan et al., 2021; Sudhan et al., 2023a, Sudhan et al. 2023b). Electromyographic (EMG) analysis plays a critical role in sports biomechanics and exercise physiology which helps in assessing muscle activity during inclined running, especially in synergistic muscles such as the GM and hamstring groups (Lewis et al., 2009). Effective hip mobility and proper GM muscle activation are necessary for optimal force output in a variety of sports dynamic movements, such as resistance training, running, sprinting, and leaping (Doud & Walsh, 1995; Van Gent et al., 2007).

Understanding how exhaustion affects muscle activation is essential to preventing injuries, particularly while running uphill. By filling in current knowledge gaps, utilizing traits such as average EMG (AEMG) signals provides important insights into how muscle activity varies with different speeds and tiredness levels, with the goal of improving sports training and injury prevention tactics.

2 Review of Literature

(1) *Masamichi Okudaira et al.* (2021) used nine male college sprinters on a motorized treadmill running 7.5 m/s to examine the impact of slope on lower extremity muscle activation during sprinting. Using surface electromyographic to measure muscle activity in the rectus femoris, vastus lateralis, biceps femoris, gluteus maximus, gluteus medius, and lateral gastrocnemius lateralis, the study examined the level and 5-degree uphill situations. Notably, during the uphill sprint's stance phase, all muscles, excluding the vastus lateralis and rectus femoris, displayed heightened activity, with an increase in rectus femoris activity during the recovery phase. These adaptations correlated with a higher step

frequency and shorter step length, indicating substantial neuromuscular adjustments in response to uphill sprinting demands. The study concluded that gluteus maximus (Gmax), gluteus medius (Gmed), biceps femoris (BF), and lateral gastrocnemius (GA) exhibited significantly increased muscle activity during the stance phase in uphill sprinting, while no statistical differences were observed in vastus lateralis (VL) between conditions.

(2) In a musculoskeletal model study, six male distance runners were investigated for lower limb muscle activities during level running (LR) and uphill running (UR) (Toshiharu Yokozawa et al., 2007). The participants engaged in overground running at various speeds on both a level runway and a 9.1% grade slope. Key findings revealed heightened activation and muscle torque in hip joint muscles (hamstrings, iliopsoas) during the recovery phase of uphill running at high speeds. Uphill running consistently showed reduced knee extension torque by the vastus during the support phase across all speeds. Additionally, the rectus femoris exhibited increased hip flexion and knee extension torques during uphill running at all speeds, indicating compensatory mechanisms to maintain a forward-leaning trunk. The study's conclusion emphasized that uphill running at high speeds amplified muscle torque in the gluteus maximus (GMAX), hamstrings (HAMS), iliopsoas (ILP), and adductors (ADD) during the recovery phase, contributing to rapid forward and backward swings of the recovery leg and an increased step frequency.

(3) *Swanson and Caldwell* (2000) conducted an integrated biomechanical analysis of high-speed incline and level treadmill running, exploring high-speed incline treadmill running (INC) at 4.5 m/s with a 30% grade, and comparing it to level running at the same speed (LSS) and stride frequency (LSSF). The study focused on joint kinematics, EMG, and swing-phase kinetics. Results revealed significantly increased stride frequency and stance phase duration during INC compared to LSS. The GM displayed normal activation before foot strike, but GM muscle activity was higher in INC post-foot strike. The study suggested that INC might provide heightened muscular loading, particularly in the gluteus maximus, making it a promising approach for sprint training to enhance forward running speed.

(4) *Sloniger Mark et al.* (1997) investigated lower extremity muscle activation during both uphill and horizontal running. Using exercise-induced contrast shifts in magnetic resonance imaging (MRI), they analyzed muscle activation patterns in 12 young women engaged in high-intensity running. Uphill running significantly increased muscle activation in the adductors (83 ± 8%), biceps femoris (79 ± 7%), and the gluteal group (79 ± 11%). In contrast to horizontal running, uphill demands led to reduced activation of the rectus femoris (29%), gracilis (18%), and semitendinosus (17%) while showing greater activation of the vastus group (23%) and soleus (14%). The study's findings in incremental cycling revealed diverse activation patterns in force-producing muscles, with the gluteus maximus demonstrating the highest relative activation at high intensities.

3 Methods

3.1 Subjects

Twenty-one male athletes, aged 24.09 ± 4.60 years, with a height of 170.45 ± 7.58 cm, weight of 64.18 ± 12.55, and BMI of 22.09 ± 4.22 kg (Table. 1), and possessing 4 ± 3.5 years of training experience, participated in this chapter. The cohort included sprinters, middle-distance runners, and athletes from soccer, handball, or basketball, engaging in a minimum of four weekly practice sessions. In order to meet the inclusion criteria, subjects had to be between 21 and 25 years of age and actively participate in state-level competitions with the right leg designated as the power leg. All participants had no lower extremity or lower back injuries in the preceding 6 months and were in good health and we have collected EMG from belly of GM (Fig.1).

Prior to data collection, participants gave their informed consent and were told not to train their lower extremities for 48 h before the test. The Institutional Review

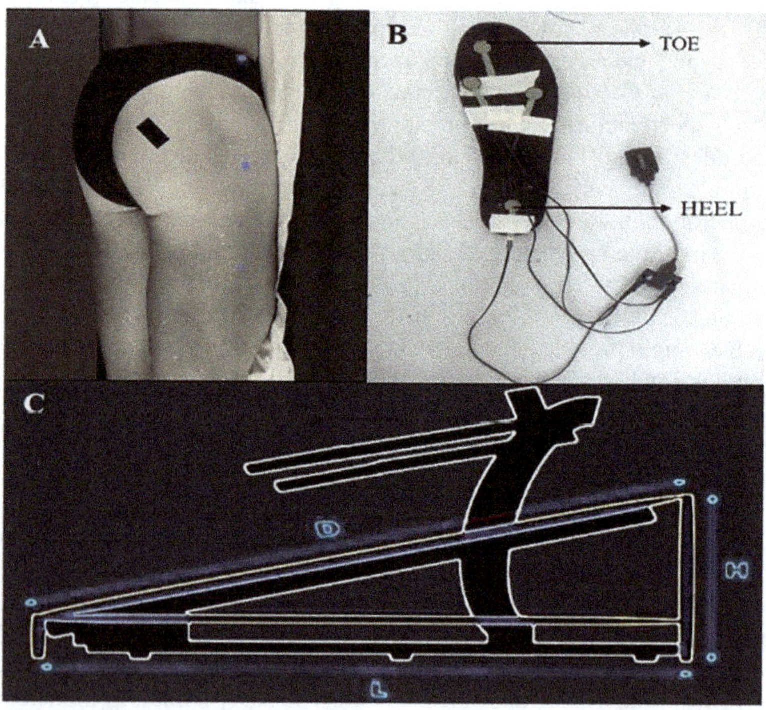

Fig. 1 (**a**) Illustration of surface EMG electrode placement on the GM. The black rectangle represents the electrode placement, while the blue-marked areas indicate the regions designated for the placement of retroreflective markers. (**b**) Optimal FSR placement on inner sole (**c**) Schematic representation of the inclined gradient setting

Table 1 Mean standard deviation of demographic data of weight, height, and BMI of 11 athletes

	N	Mean	Std. deviation
Age	21	24.09 ± 4.60	0.83
Weight	21	64.18 ± 12.55	7.36
Height	21	170.45 ± 7.58	2.98
BMI	21	22.09 ± 4.22	2.48

Board has granted approval for the use of human participants, guaranteeing that scientific research adheres to ethical guidelines.

3.2 Data Collection

Surface electromyographic (EMG) electrodes were accurately positioned on the GM muscle belly of each participant through manual muscle testing. A modified motion sensor marker set (Kadaba et al., 1990) facilitated on the bilateral placement of retroreflective markers on the lower extremities, lateral femoral epicondyle, lateral greater trochanter of the hip, lateral shank, and lateral malleolus (Dadfar et al., 2021) and an additional marker on the lateral femur. The determination of leg dominance was based on the ball-kicking leg.

Participants self-reported engaging in regular running at speeds exceeding 24 km/h, covering an average distance of 30 ± 12 km per week in the 3 months leading up to the chapter. The treadmill gradient (slope) was checked for inclined running 1 day before testing. Subjects underwent orientation regarding the chapter's purpose, followed by a 12-min general warm-up comprising dynamic stretches for the lower extremities. The warm up involved brisk walking at 3.5 km/h with a 0% gradient, jogging at 4–5 km/h with a 4% gradient, and jogging at 5–6 km/h with a 5% gradient, in addition to a self-paced warm-up on the treadmill.

Adherence to SENIAM guidelines (Hermens et al., n.d.) for appropriate clothing ensured access to EMG electrode placement sites and refer to Atlas of muscle innervation zones in different body parts (Barbero et al., 2012). Skin preparation involved hair removal, cleaning, and abrasion with an isopropyl alcohol swab. The electrodes were situated distally to the motor point of each muscle, and amplification was fine-tuned for optimal resolution through the utilization of pre-amplified Ag/AgCl surface electrodes (Delsys Inc., Boston, MA, USA). Muscle activity was detected using the eight-channel Trigno Avanti EMG sensors, which were supplied by AD Instruments (Delsys Inc., Boston, MA, USA) and interfaced with an amplifier/processor module. Rectangular electrodes (37 × 26 × 15mm) were placed on the left and right GM. Participants completed seven inclined running trials with their maximum effort, each separated by 1 min of rest. Successful EMG and kinematic data collection defined an acceptable trial (Hanon et al., 2005).

Motion analysis videography has been done with BaSix and an analog data acquisition system (Motion Analysis Inc., cortex, USA). Retroreflective marker trajectories were recorded at a sample rate of 240 frames per second. The EMG signal collection was synchronized with the marker trajectories and recorded at a rate of 2400 samples per second using the (Delsys Inc., Boston, MA, USA) EMG system. The entire data collection process adhered to time synchronization.

3.3 Data Reduction

In each trial condition, participants initially ran on a treadmill with a 12% gradient (Sinclair et al., 2013) and a speed of 7 km/h for 2 min, allowing for electrode and frequency checks within proper activation ranges. Once comfortable with the inclined configuration, subjects were then instructed to run for 12 min on the treadmill, varying speeds from 7.5 km/h to 16 km/h with a 15% gradient.

During the running session, signals were recorded using Ag/AgCl electrodes, chosen for their nonpolarizable nature, providing stable electrode-skin interfaces. The study's electrodes showed resistance instead of capacitance, meaning that changes in the distance between the skin and the electrode surface had less effect on how well they worked. These electrodes were placed along the longitudinal midline of the GM muscles, between two motor sites, and between motor points and tendon insertions. By ensuring a steady electrode-skin interaction, this positioning lessened interference, and noise was clearly monitored. Participants initiated running at a 15% gradient with a speed of 13.5 km/h for 1 min, followed by a 1-min recovery period after each repetition. Subsequently, subjects continuously ran with a 15% incline, adjusting treadmill speed incrementally to maintain consistent stride frequency.

After a 1-min rest period, subjects completed six trials, each lasting 45 s to 1 min, with speed variations ranging from 13.5 km/h to 16 km/h. The entire process was conducted under inclined conditions on the treadmill, with 1-min rests between each trial. EMG data from the left gluteus maximus (L-GM) and right gluteus maximus (R-GM) maximus of the lower extremities were collected for further analysis, encompassing a 200-meter run (approximately 45 s) and a 1-min recovery, over 6 repetitions.

The time it takes for a full cycle of a foot's movement, from heel touchdown to the next similar gate cycle, is measured along with the time the foot spends away from the ground (from toe-off to heel contact) and the time it takes for the foot to make contact with the ground (heel touchdown). Additionally, each runner's shoe has an insole with four FSR sensors that record biometric data. The wires and sensor stripes were secured using medical tape. This particular insole is designed with four pressure-sensitive zones, and the analysis concentrates on the zones corresponding to the heel and toe. Additionally, the investigation examined pressure distribution and foot placement at the first and fifth metatarsal areas.

The 3D coordinates of reflective markers were filtered using a fourth-order Butterworth digital filter for better accuracy (Yu et al., 1997). The coordinates of specific anatomical landmarks were estimated, and the whole body using a different body segment model (Sinclair et al., 2013). Reference frames were established for the pelvic and thigh segments, and the 3D coordinates of GM attachment sites were determined by employing relevant muscle attachment points within the tibia and pelvis segment reference frames. Raw EMG signals underwent filtering processes, including band-pass and rectification, followed by low-pass filtering to obtain linear envelope EMG data. This comprehensive methodology ensured accurate and synchronized data collection for subsequent analysis.

3.4 Data Analysis

The properties of the contractions of muscles in the GM on both lower extremities were evaluated using two independent 2 (muscle) × 6 (repetition) analyses of variance using a mixed design. The analyses focused on three dependent variables: (1) maximum muscle contraction during each repetition phase measured by EMG, (2) the magnitude of EMG contraction speed, and (3) the EMG magnitude and force distribution.

The analysis examined the GM muscle during both the stance and repetition phases. Repetition was treated as a repeated measure (RM ANOVA), while muscle was considered independent Post hoc t-tests with Bonferroni adjustment were conducted to identify any significant differences when a main effect was observed (Yu et al., 2008). Unless otherwise stated in the data, all significant differences are reported at a level of $P < 0.05$. Data were acquired using LabChart Reader 8.0, and the conversion to RMS value in the software was facilitated. The Rcode and RStudio computer program package for statistical analysis was utilized for all statistical analyses.

4 Result

Across the stride period, flight period, and contact period, all exhibited significant reductions across the speed range, decreasing from the initial velocity (13.5 km/h) to the test's completion (16 km/h), with values decreasing from 554 to 494 ms, 345 to 309 ms, and 263 to 185 ms, respectively ($P < 0.05$).

Stride duration differed significantly at intermediate speeds (14.5 and 15 km/h), and contact duration differed at the end (15.5–16 km/h). Flight duration exhibited a significant difference as well. Table 2 illustrates that stride parameter durations remained consistent during the initial three repetitions and began to decrease significantly after the third repetition. A noticeable decrease was observed in the sixth repetition, as compared to the third repetition, in both flight phase and stride duration. The duration of the contact phase remained stable throughout the test, except for the final repetition, which exhibited a significant difference.

Table 2 Exploring the impact of speed and stage duration on stride parameters across repetitions R1–R6

	Repetition	R1	R2	R3	R4	R5	R6
	speed (km/h)	13.5	14	14.5	15	15.5	16
Stride cycle duration (ms)	M	554	574	593	540	505	494[a]
	Ds	22	26	29	21	19	14[a]
Flight duration (ms)	M	345	332	330	317	298	309[a]
	Ds	14	13	11	15	17	17[a]
Contact duration (ms)	M	209	242	263	223	207	185[a]
	ds	9	6	5	9	9	14[a]

[a]$P < 0.05$ Significant difference between the third and sixth repetition on inclined running

Evaluation of EMG parameters involved the assessment of the GM on both lower extremities. Figure 1 illustrates the placement of EMG electrodes, the measurement of the inclined gradient, and the distance scale (blue-covered line) for muscle activation during each repetition within the 45–60-second running cycle. The muscular activation patterns for the L-GM and R-GM manifested as a continuous burst of activation. Given the challenging nature of dividing this burst into distinct segments, we treated it as a unified activation episode.

Muscle Activation Pattern

The duration of muscle activation patterns was assessed, and measurements were expressed in microvolts (μV) or the duration of each stride repetition was measured relative to the methods section. When converted to millivolts, activation durations varied significantly among participants, and the averages were computed for the final results. For instance, at a speed of 13.5 km/h, the burst of activation in the L-GM during the first repetition (R) phase (L-GM R1) increased from 0.113 ± 0.02 mV to 0.137 ± 0.16 mV in the sixth repetition (L-GM R6).

When considering the burst durations relative to the stride duration, there were notable differences between repetitions in the muscular amplitudes (Inacio et al. 2014), particularly at 13.5 km/h. In the L-GM, motor units were retained, and activation increased gradually from R1 (0.113 ± 0.02) to R6 (0.137 ± 0.16), indicating increased recruitment. In contrast, the R-GM exhibited a deviation in activation, with maximal activation occurring in the second repetition. Subsequently, activation gradually decreased followed by a decline from the third to the sixth repetition (Table 3), indicating fatigue progression in the R-GM muscle. Notably, R-GM showed more variation compared to L-GM. Despite a gradual decrease in action potential in the fifth repetition, R-GM retained the amplitude gradually in the last repetition. R2 had the highest muscle activation (average of 0.096 mV), while R3, R4, and R5 had the same average frequency of 0.072 mV. Overall, R-GM exhibited the least activation among the hip extensors compared to L-GM.

The impact of increasing speed on muscle activation patterns varied significantly. The study found that the length of activation in the R-GM decreased significantly ($P < 0.05$) as speed increased, while the L-GM increased significantly. During inclined running on the treadmill, the gluteus maximus muscles are responsible for

Table 3 Root mean square (RMS) of muscle activation in R-GM and L-GM muscles across six repetitions of inclined running

Repetition	LGlut	RGlut
1	0.113 ± 0.02	0.093 ± 0.06
2	0.116 ± 0.02	0.096 ± 0.05
3	0.111 ± 0.03	0.070 ± 0.02
4	0.112 ± 0.04	0.071 ± 0.02
5	0.128 ± 0.11	0.074 ± 0.02
6	0.137 ± 0.16	0.065 ± 0.02

hip extension, involving the backward movement of the thigh, where the activation period was longer relative to stride duration as speed increased from the beginning to the end of the test.

5 Discussion

The main findings in this chapter reveal that, in the context of inclined running, the GM, a hip extensor muscle, undergoes fatigue due to increased use at higher speeds. Notably, the dominant leg's R-GM exhibits earlier signs of fatigue compared to other examined lower limb muscles. Conversely, the L-GM muscles show heightened activation levels in response to speed increments. Additionally, the most effective method for assessing muscular fatigue during treadmill running involves the work done by the intensity of training sessions by EMG analysis using a distance scale and normalized muscle activation patterns in line with existing literature (Franz & Kram, 2012). Rehabilitation focusing on strengthening both GM muscles contributes to a balanced hip extension during running, preventing injuries and enhancing running gait and speed. Including inclined running in rehabilitation targets explicitly the lower extremities, particularly the GM muscle, promoting greater strength and adaptability for improved running performance by various speed training plans and adaptation through training periodically.

Analytical Approaches for Data Analysis

The literature provides various analytical approaches to EMG analysis using spectral density, focusing on the onset of fatigue during inclined treadmill running. This topic has been explored by Giandolini et al. (2016), Mizrahi and Verbitsky (1997), and Vernillo et al. (2015). A commonly employed method involves testing the development of integrated electromyographic (iEMG) across a predetermined number of bursts, a technique proposed by researchers such as Hanon et al. (1998), Taylor and Bronks (1995), and Piucco et al. (2020). This analytical strategy aims to investigate and comprehend EMG activity patterns as indicators of muscle exhaustion, particularly in jogging, running, or sprinting on an inclined treadmill.

The current methodologies for analyzing running characteristics have limitations. Notably, in prolonged treadmill sessions at a constant speed, an increase in stride frequency while maintaining the last repetition of the test has been observed, a phenomenon noted in previous studies (Bourne et al. 2021). Consequently, variations in the characteristics of each activation burst may occur concurrently with changes in both stride length and frequency. Our investigation noted increased stride frequency with fatigue, leading to a reduced covered distance for a set number of bursts but potentially higher when considering a fixed duration. Additionally, the activation pattern exhibited variations with increasing repetitions. The choice of the analytical method becomes crucial, revealing disparities in covered distance and work accomplished. With a fixed speed and increased frequency, burst analysis emerges as a stable parameter, resulting in decreased stride length and the number of impulses during the contact phase within a given stride cycle (Mero et al. 1992). Noteworthy, the onset of fatigue, characterized by an increase in EMG during constant submaximal contractions (Enoka & Stuart, 1992), may be compensated for and potentially masked by this reduction in the number of impulses.

In addressing about the methodological challenge, utilizing the average activation level (activation level divided by the duration of activation) does not provide a viable solution. A more effective approach involves normalization with maximum voluntary isometric contraction (MVIC) percentages. During inclined running, the activation of the L-GM muscle varied between 57% and 65%, reaching its peak in the sixth repetition at 64.62% (refer to Fig. 2). Notably, the R-GM exhibited the lowest activation, constituting 35.38% over the six repetitions, indicating a significant disparity in muscle activation (Table 2). The mean standard deviation ranged from 0.093 ± 0.06 to 0.065 ± 0.02 mV throughout inclined running.

Particularly noteworthy were the muscles with the most pronounced activity, notably the L-GM, which recorded 0.137 ± 0.16 mV during the sixth repetition. In inclined conditions, it became apparent that the L-GM muscle retained motor units, with activation progressing from 0.113 ± 0.02 in repetition 1 to 0.137 ± 0.16 in repetition 6. Importantly, a correlation with a p-value of less than 0.5 was observed in the percentage activation results for the left and right gluteus muscles. The extent and duration of activation are influenced by the combined impact of heightened frequency and fatigue. It is worth noting that when muscles become tired, their activation levels may either increase or stay the same, depending on the specific muscle. This suggests that fatigue can have a greater impact on the activation level than frequency increase. In addition, the duration of bursts may show different patterns of change in tired muscles, such as remaining constant in the R-GM, decreasing in the R-GM, or increasing in the L-GM (as shown in Fig. 3). Therefore, normalized mean muscle activation is only relevant if there is no variation in biomechanical parameters.

The importance of the gluteus maximus (GM) and hip extensors in sprinting is well-documented. Neumann (2010) highlights the crucial role of the GM, showing a significant increase in electromyographic (EMG) activity as running speed increases (Lieberman et al., 2006; Sloniger et al., 1997). In our chapter, the GM exhibited an average activation level of 65% of the normalized peak EMG, occurring at 64.62% muscle activation of the stride cycle (at a speed of 16 km/h).

The early fatigue observed in the L-GM from the first to the second repetition in our investigation can be attributed to the heightened activation of L-GM after the

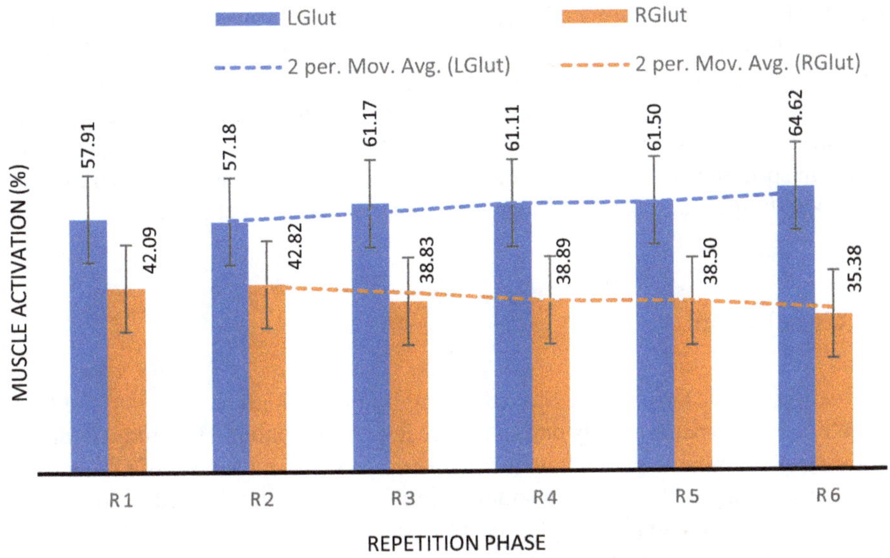

Fig. 2 Averaged normalized muscle activation of left and right gluteus during six repetitions of inclined running, presented as a percentage of maximum voluntary isometric contraction (MVIC) *Note: A trend line representing a two-period moving average is applied from the second to the sixth repetition for enhanced trend analysis. Blue dash for L-GM and orange for R-GM

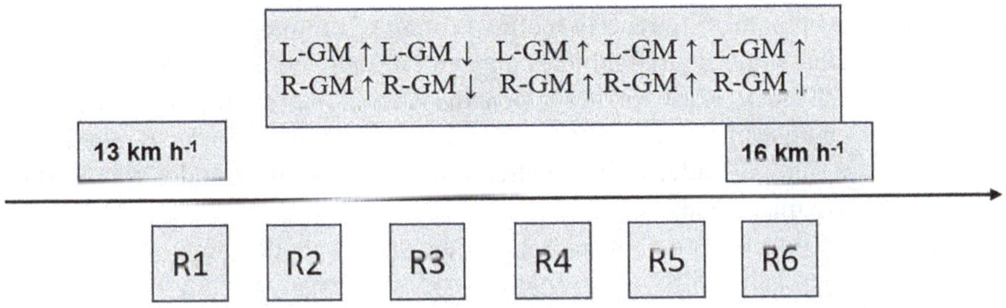

Fig. 3 Sequential emergence of fatigue thresholds indicated by surface EMG in both the L-GM and R-GM throughout all repetitions. ↑ represents increment in amplitude and ↓ decrement in amplitude

second repetition, particularly as the running speed increased. The early fatigue in the L-GM raises questions; despite being recruited for a comparable duration (58% of the time at 13.5 km/h to 14 km/h), the L-GM exhibited higher activation levels than the R-GM, which registered at 35.38% of normalized EMG. One plausible explanation is the differential degree of knee flexion. Greater hip flexion may reduce work done by hip extensors during fatigue (Hip Extension for Efficient Running—Gregor Mc, J Stephen, 2010; Yokozawa et al., 2007).

In the latter phases of repetition, there was a rise in activation of the L-GM, potentially serving as compensation for the muscle fatigue observed in the

R-GM. This illustrates intricate interactions among muscle groups and the dynamic adjustments required to sustain performance across different conditions and lateral variation of activation in a complete gait cycle of inclined running. As elaborated in the discussion section, these findings offer valuable insights for refining training methods by focusing on particular muscles that help to develop and enhance performance in sprinting athletes by elucidating the intricate relationship between muscle activation, fatigue, and compensatory mechanisms (Okudaira et al., 2021).

6 Conclusion

To the best of our knowledge, limited research has explicitly delved into the GM muscle activation pattern study during incline treadmill running. This study uniquely addresses this research gap by conducting a comprehensive analysis of GM activation and study about the variation and utilizing force estimates derived from surface electromyographic signals. Employing FSR sensors, the research quantifies force proportionate to GM muscle activation during specific hip extension's dynamic movements in inclined running gait, encompassing both the L-GM and R-GM muscles. Force estimation is meticulously studied and implemented based on the measured EMG signals to understand the kinematic and kinetic analysis. The obtained results reveal a noteworthy trend: in right-dominant athletes, the activation of the left L-GM increases, while the amplitude of the R-GM decreases during overuse, contributing to immediate fatigue in inclined running. From a physiological standpoint, our chapter suggests that athletes engaged in inclined/uphill speed development during the back phase of hip extension should incorporate eccentric and concentric strength training for balanced GM activation. Failure to do so may result in variations in running patterns due to altered muscle activation, and it may lead to major injury like muscle tear.

One potential limitation of our chapter is the small number of subjects, which may limit the generalizability of muscle-dependent patterns. Future work aims to address this limitation by continually recruiting more volunteers to expand our database. Additionally, we plan to optimize our coding to enable the simultaneous collection of kinetic, kinematic, and EMG data in a single stretch, moving away from using pre-recorded EMG signals. This enhancement will provide a more comprehensive and real-time understanding of the complex interplay between muscle activation and biomechanics during inclined running.

Acknowledgments We thank Vipin Venugopal for his valuable recommendations on the biomedical approach and insightful discussions. Special thanks are extended to Bharanidaran for his support in conducting statistical analyses and to Balamurugan for his valuable contributions to data processing.

Conflicts of Interest The authors declare there is no conflict of interest.

References

Barbero, M., Merletti, R., & Rainoldi, A. (2012). Atlas of muscle innervation zones. In *Atlas of muscle innervation zones*. Springer Milan. https://doi.org/10.1007/978-88-470-2463-2

Bourne, M. N., Pollard, C., Messer, D., Timmins, R. G., Opar, D. A., Williams, M. D., & Shield, A. J. (2021). Hamstring and gluteal activation during high-speed overground running: Impact of prior strain injury. *Journal of Sports Sciences, 39*(18), 2073–2079. https://doi.org/10.1080/02640414.2021.1917839

Buckthorpe, M., Stride, M., & Villa, F. D. (2019). Assessing and treating gluteus maximus weakness—A clinical commentary. *International Journal of Sports Physical Therapy, 14*(4), 655–669. https://doi.org/10.26603/ijspt20190655

Dadfar, M., Sheikhhoseini, R., Jafarian, M., & Esmaeili, A. (2021). Lower extremity kinematic coupling during single and double leg landing and gait in female junior athletes with dynamic knee valgus. *BMC Sports Science, Medicine and Rehabilitation, 13*(1), 152. https://doi.org/10.1186/s13102-021-00385-y

Doud, J. R., & Walsh, J. M. (1995). Muscle fatigue and muscle length interaction: effect on the EMG frequency components. *Electromyography and Clinical Neurophysiology, 35*(6), 331–339.

Eliot Hu, Y.-W., & Ho, G. W. (n.d.). *Deep gluteal syndrome: A pain in the buttock*. http://journals.lww.com/acsm-csmr

Enoka, R. M., & Stuart, D. G. (1992). Neurobiology of muscle fatigue. *Journal of Applied Physiology, 72*(5), 1631–1648. https://doi.org/10.1152/jappl.1992.72.5.1631

Fiore, J. (2017). Hip extension for efficient running—The runners edge. *Physical Therapy*. https://www.runnersedgemt.com/2017/hip-extension-for-efficient-running/

Franettovich Smith, M. M., Bonacci, J., Mendis, M. D., Christie, C., Rotstein, A., & Hides, J. A. (2017). Gluteus medius activation during running is a risk factor for season hamstring injuries in elite footballers. *Journal of Science and Medicine in Sport, 20*(2), 159–163. https://doi.org/10.1016/j.jsams.2016.07.004

Franz, J. R., & Kram, R. (2012). The effects of grade and speed on leg muscle activations during walking. *Gait & Posture, 35*(1), 143–147. https://doi.org/10.1016/j.gaitpost.2011.08.025

Gregor Mc, J Stephen (2010). The Runner's Edge High-Tech Training for Peak Performance. Stephen J, Human Kinetics, 107–120.

Giandolini, M., Vernillo, G., Samozino, P., Horvais, N., Edwards, W. B., Morin, J.-B., & Millet, G. Y. (2016). Fatigue associated with prolonged graded running. *European Journal of Applied Physiology, 116*(10), 1859–1873. https://doi.org/10.1007/s00421-016-3437-4

Hanon, C., Thépaut-Mathieu, C., Hausswirth, C., Le Chevalier, J. M., Le, J. M., Electromyogram, C., Hanon, C., Theâ Paut-Mathieu, C., & Hausswirth, C. (1998). Electromyogram as an indicator of neuromuscular fatigue during incremental exercise. *European Journal of Applied Physiology, 78*(4), 315. https://doi.org/10.1007/s00421-005-0426-1

Hanon, C., Thépaut Mathieu, C., & Vandewalle, H. (2005). Determination of muscular fatigue in elite runners. *European Journal of Applied Physiology, 94*(1–2), 118–125. https://doi.org/10.1007/s00421-004-1276-1

Hawkins, R. D., Hulse, M. A., Wilkinson, C., Hodson, A., & Gibson, M. (2001). *The association football medical research programme: an audit of injuries in professional football*. www.bjsportmed.com

Hermens, H. J., Freriks, B., Merletti, R., Stegeman, D., Blok, J., Rau, G., Disselhorst-Klug, C., & Hägg, G. (n.d.). *European recommendations for surface electromyography results of the SENIAM project*.

Inacio, M., Ryan, A. S., Bair, W.-N., Prettyman, M., Beamer, B. A., & Rogers, M. W. (2014). Gluteal muscle composition differentiates fallers from non-fallers in community dwelling older adults. *BMC Geriatrics, 14*(1), 37. https://doi.org/10.1186/1471-2318-14-37

Jagadevan, M., Mohanakrishnan, B., Bhavanani, A. B., Shristhudhi, D., Arumugam, P., Subbiah, B., Chandrashekhar, K., & Swaminathan, P. (2021). Additive effect of "Brahma Mudra" on pain, proprioception and functional abilities in non-specific mechanical neck pain. *Journal of Bodywork and Movement Therapies, 27*, 717–722. https://doi.org/10.1016/j.jbmt.2021.06.015

Jeong, U.-C., Sim, J., Kim, C.-Y., Hwang-Bo, Gak, & Nam, C. (n.d.). The effects of gluteus muscle strengthening exercise and lumbar stabilization exercise on lumbar muscle strength and balance in chronic low back pain patients. *Journal of Physical Therapy Science*.

Kadaba, M. P., Ramakrishnan, H. K., & Wootten, M. E. (1990). Measurement of lower extremity kinematics during level walking. *Journal of Orthopaedic Research, 8*(3), 383–392. https://doi.org/10.1002/jor.1100080310

Kaiyaperumal, A. B., Subbiah, B., Paulraj, M., & Sudhan, P. (2022). Prevention Enhance Program (Pep) with proprioceptive training on the recurrence of ACL injury for post-ACL reconstruction among football players. *Neuroquantology, 20*(19), 143–148.

Kaiyaperumal, A. B., Subbiah, B., & Paulraj, M. (2023). The multimodal exercises on physical, functional, agility, and perturbation for post-ACL reconstruction among athlete. *International Journal of Life Science and Pharma Research*, 175–185.

Kara, M., Tiftik, T., Özcan, H. N., Kaymak, B., & Özçakar, L. (2015). Gluteus Maximus muscle tear as a rare cause of hip pain and sciatica. *Journal of Emergency Medicine, 49*(5), 705–706. https://doi.org/10.1016/j.jemermed.2015.01.040

Lewis, C. L., Sahrmann, S. A., & Moran, D. W. (2009). Effect of position and alteration in synergist muscle force contribution on hip forces when performing hip strengthening exercises. *Clinical biomechanics, 24*(1), 35–42. https://doi.org/10.1016/j.clinbiomech.2008.09.006

Lieberman, D. E., Raichlen, D. A., Pontzer, H., Bramble, D. M., & Cutright-Smith, E. (2006). The human gluteus maximus and its role in running. *Journal of Experimental Biology, 209*(11), 2143–2155. https://doi.org/10.1242/jeb.02255

Manickavelu, P. S. B., & Kaliyaperumal, A. B. (2022). Prevalence of diurnal physical mobility and sedentary behavior among allied healthcare college students in Puducherry, India. *International Journal of Epidemiology and Health Sciences, 3*(3), 1. https://doi.org/10.51757/IJEHS.3.3.2022.249173

Manickavelu, P., Subbiah, B., Kaliyaperumal, A. B., & Benjamin, S. D. (2022). The relationship between daily physical activity and respiratory function, flexibility, and body composition among allied healthcare college students. *NeuroQuantology, 20*(16), 1620–1629.

Mero, A., Komi, P. V., & Gregor, R. J. (1992). Biomechanics of sprint running. *Sports Medicine, 13*(6), 376–392. https://doi.org/10.2165/00007256-199213060-00002

Miller, R., Balshaw, T. G., Massey, G. J., Maeo, S., Lanza, M. B., Haug, B., Johnston, M., Allen, S. J., & Folland, J. P. (2022). The muscle morphology of elite female sprint running. *Medicine & Science in Sports & Exercise, 54*(12), 2138–2148. https://doi.org/10.1249/MSS.0000000000002999

Mizrahi, J., & Verbitsky, O. (1997). The influence of fatigue on EMG and impact acceleration in running. https://www.researchgate.net/publication/286306578

Neumann, D. A. (2010). Kinesiology of the hip: a focus on muscular actions. *Journal of Orthopaedic & Sports Physical Therapy, 40*(2), 82–94. https://doi.org/10.2519/jospt.2010.3025

Nunes, G. S., Pizzari, T., Neate, R., Barton, C. J., & Semciw, A. (2020). Gluteal muscle activity during running in asymptomatic people. *Gait and Posture, 80*, 268–273. https://doi.org/10.1016/j.gaitpost.2020.06.008

Okudaira, M., Willwacher, S., Kuki, S., Yoshida, T., Maemura, H., & Tanigawa, S. (2021). Effect of incline on lower extremity muscle activity during sprinting. *The Journal of Physical Fitness and Sports Medicine, 10*(2), 67–74. https://doi.org/10.7600/jpfsm.10.67

Piucco, T., Diefenthaeler, F., Prosser, A., & Bini, R. (2020). Validity of different EMG analysis methods to identify aerobic and anaerobic thresholds in speed skaters. *Journal of Electromyography and Kinesiology, 52*, 102425. https://doi.org/10.1016/j.jelekin.2020.102425

Sinclair, J., Richards, J., Taylor, P. J., Edmundson, C. J., Brooks, D., & Hobbs, S. J. (2013). Three-dimensional kinematic comparison of treadmill and overground running. *Sports Biomechanics, 12*(3), 272–282. https://doi.org/10.1080/14763141.2012.759614

Sloniger, M. A., Cureton, K. J., Prior, B. M., & Evans, E. M. (1997). Lower extremity muscle activation during horizontal and uphill running. *Journal of Applied Physiology, 83*(6), 2073–2079. https://doi.org/10.1152/jappl.1997.83.6.2073

Sudhan, P., Ananthan, B., Janaki, G., Rajagopalan, N., Subbiah, B., & Sukumaran, R. (2023a). Effect of yoga therapy on neurological characteristics in diabetic peripheral neuropathy: Neuro health perspective. *Journal for ReAttach Therapy and Developmental Diversities, 10*(2), 1071–1078.

Sudhan, P., Subbiah, B., Rajagopalan, N., Parveen, J., Sukumaran, R., & Adnanthan, J. G. (2023b). The neurological features of diabetic peripheral neuropathy (DPN) patients are addressed by applying varma therapy to biochemical alterations. *European Chemical Bulletin, 12*(10), 240–251.

Swanson, S. C., & Caldwell, G. E. (2000). An integrated biomechanical analysis of high speed incline and level treadmill running. *Medicine & Science in Sports & Exercise, 32*(6), 1146–1155. https://doi.org/10.1097/00005768-200006000-00018

Taylor, A. D., & Bronks, R. (1995). Reproducibility and validity of the quadriceps muscle integrated electromyogram threshold during incremental cycle ergometry. *European Journal of Applied Physiology and Occupational Physiology, 70*(3), 252–257. https://doi.org/10.1007/BF00238572

Van Gent, R. N., Siem, D., Van Middelkoop, M., Van Os, A. G., Bierma-Zeinstra, S. M. A., & Koes, B. W. (2007). Incidence and determinants of lower extremity running injuries in long distance runners: A systematic review. In. *British Journal of Sports Medicine, 41*(8), 469–480. https://doi.org/10.1136/bjsm.2006.033548

Vasudevan, J. M., Smuck, M., & Fredericson, M. (2012). *Evaluation of the athlete with buttock pain*. www.acsm-csmr.org

Vernillo, G., Savoldelli, A., Zignoli, A., Skafidas, S., Fornasiero, A., La Torre, A., Bortolan, L., Pellegrini, B., & Schena, F. (2015). Energy cost and kinematics of level, uphill and downhill running: Fatigue-induced changes after a mountain ultramarathon. *Journal of Sports Sciences, 33*(19), 1998–2005. https://doi.org/10.1080/02640414.2015.1022870

Yokozawa, T., Fujii, N., & Ae, M. (2007). Muscle activities of the lower limb during level and uphill running. *Journal of Biomechanics, 40*(15), 3467–3475. https://doi.org/10.1016/j.jbiomech.2007.05.028

Yu, B., Stuart, M., Kienbacher, T., Growney, E., & An, K.-N. (1997). Valgus-varus motion of the knee in normal level walking and stair climbing. *Clinical biomechanics, 12*(5), 286–293. https://doi.org/10.1016/S0268-0033(97)00005-3

Yu, B., Queen, R. M., Abbey, A. N., Liu, Y., Moorman, C. T., & Garrett, W. E. (2008). Hamstring muscle kinematics and activation during overground sprinting. *Journal of Biomechanics, 41*(15), 3121–3126. https://doi.org/10.1016/j.jbiomech.2008.09.005

Analysis of Research Trends in Sports Analytics for Soccer from 2012 to 2023

Ravinder Kumar Verma and Manish Kumar Shrivastava

1 Introduction

The emerging technologies support many countries to enable data-driven decision-making and significantly contribute to multiple domains. There has been significant research on the use of analytics in sports over the past several years (Bock, 2017; Bonomo et al., 2014; de Jong et al., 2022; Seakins et al., 2023). Sports analytics research provides various insights into the application of decision science approaches to sports, including predictive analytics of sports events, impacting real-time sports situations, improving the probability of a win, and supporting the pre- and post-analysis of sports events (Alarcón et al., 2017; Bauer & Anzer, 2021; de Jong et al., 2022; Mavrogiannis & Maglogiannis, 2022; Vinué & Epifanio, 2017). The lack of data influences the planning that will fail the sports management to improve, organize, and plan in the short and long term. The lack of data and poor sports data management lead to ambiguity, which may have significant consequences. Primarily developed countries invest in sports analytics to improve the game by improving preparation and game strategies (Ács et al., 2023). Academic research focuses on sports analytics, especially after the emergence of analytics tools in the last decade. With sports' immense popularity, they have enabled the stakeholders to find means to improve the game and increase its revenue (Ács et al., 2023; Matano et al., 2023). To efficiently manage the game, many developed countries have adopted sports analytics (SA) as a strategy. However, the sports analytics implementation process is very complex, particularly when independent variables are not fixed due to contextual and individual factors. Therefore, the primary motivation to use sports analytics is adoption by a few well-established sports agencies (Alarcón et al., 2017;

R. K. Verma (✉) · M. K. Shrivastava
OP Jindal Global University, Sonipat, Haryana, India

Christ University, Bangalore, Karnataka, India

A. Mansurali et al. (eds.), *Sports Analytics*,
https://doi.org/10.1007/978-3-031-63573-1_9

Bonomo et al., 2014; Fernández et al., 2021). The research indicates that the innovation ecosystems and continuous interactions between technology and sports influence the main reasons developed countries use emerging sports technologies. The traditional sports practices were quantified and converted to statistical models with multiple variables. In developed countries, sports analytics usage is more oriented towards sports like soccer (Anzer & Bauer, 2022; Ariyaratne & Silva, 2022; Freitas et al., 2021). SA has positive effects on the performance of individuals and games. SA positively impacts the soccer game by improving the players' game and game revenue and influencing the popularity of the players and games. SA provides new opportunities and resources for new players and teams to improve their performance in specific areas like measuring the response time, coordination between players, and understanding the competitor team; these aspects contribute to the game's overall performance.

Research has shown that SA can provide timely feedback about the individual game and team performance on various parameters. SA improve the performance and revenue and provide new opportunities for new players and teams to improve their game by investing in technology inputs, the process of planning a game, and adopting different strategies to create an encouraging sports environment (Borghesi et al., 2023; Dobreff et al., 2022; Pappalardo et al., 2019; Roumani, 2022). The limited resources and lack of technical support for SA and limited use of advanced means of SA in soccer sports are significant barriers to improving sports environments (Bonomo et al., 2014; Ćwiklinski et al., 2021; Rahimian et al., 2023; Roumani, 2022). The research indicated that multiple countries are researching SA applications. The gradual expansion of the SA to developing countries represents a series of encouraging effects of SA to use by countries to improve the game's current state. The use of SA led to redefining sports practices, such as adjusting to the context, strategies against individuals and teams, and changes in team composition to achieve higher performance.

This research uses bibliometric analysis for an in-depth investigation into the characteristics of published literature. Bibliometric analysis can highlight an institution's academic strength by highlighting the research output. The research impacts can be assessed through citation and co-citation, which allows the exploration and clarification of the primary work and the advancement of a particular discipline (Donthu et al., 2021). Modern bibliometric techniques consider the influence of top academics and other periodicals along with statistical computations. They strongly emphasize how knowledge created and changes in a subject significantly impact upcoming research and development. We performed the bibliometric analysis to understand research developments and trends (Donthu et al., 2020, 2021; Zupic & Čater, 2015) in the SA for soccer.

This research contributes to the following: (1) to understand the statistical features of SA in soccer research, types, patterns, and citations; (2) to present the research collaboration networks between articles and discuss the research papers from the regions, institutions, and authors; and (3) to examine keywords for co-occurrence analysis and timeline view analysis to understand the themes of SA in soccer trends.

The subsequent sections of this paper are organized in the following manner: Section 2 introduces the bibliometric approaches employed in this study and offers a comprehensive discussion on the data sources utilized for the research. Section 3 of this study presents the findings by showing the essential statistical features, cooperation networks, and keyword analysis of the research on sports analytics of football sports. In addition, we engaged in a discussion regarding the work. Section 4 presents the conclusion, research implications, and future research and limitations.

2 Methods

This research adopted the bibliometric analysis (Ninkov et al., 2022; Pritchard, 1969), which provides a scientific approach to study management research (Cuccurullo et al., 2016; Donthu et al., 2020, 2021; Ellegaard & Wallin, 2015; Maggon, 2022) and other domains. The research was conducted in three parts: first, the database was curated; second, the data was refined and retrieved; and third, a bibliometric evaluation was performed. This study used the Scopus database, commonly used in bibliometric research on sports analytics. Scopus was chosen for its extensive coverage of science, engineering, and social science disciplines and its more significant number of recently indexed journals when compared to other databases like Web of Science (Brzezinski, 2015; Donthu et al., 2021). This holds significant importance, especially in the field of analytics research. The data was collected in May 2023. We conducted a thorough literature review in the second phase using the Scopus online research database. The search field used to query the database was "Title-Abstract-Keywords". A Boolean operator was used along with the following query: We performed a thorough literature search by utilizing the search term "Sports analytics" and narrowing our search to articles that specifically mentioned "soccer" or "football" in the title or abstract. Our primary sources consisted of original research journal articles written in English. To narrow the search results, a temporal constraint was applied to the publication period, specifically from 2012 to 2023. The decision was made based on the acknowledgement that the realm of sports analytics experienced a significant increase in research endeavours starting in 2012. The investigation yielded an overall result of 279 published outputs.

We reviewed research article keywords for the analysis to ensure theme connection and remove unrelated publications (Donthu et al., 2021). "Sports Analytics", "Football", "Soccer", "Association Football", "Soccer Player", "Soccer Analytics", "Football Analytics", "Soccer Analysis", "Association Football, Performance Analysis, Soccer, Sports Analytics", "Football Game", "Football Recruiting". The research was limited to ((LANGUAGE, "English") and (LIMIT-TO (SRCTYPE, "j")). Both co-authors independently wrote abstracts. After applying the filters described above, we received 142 research works. After combining and merging the independent screenings, we excluded 137 publications and preserved 142 for our bibliometric study. Bibliometric analysis was the third step. Because of its practical workflow-based bibliometric analysis, we selected "Bibliometrix", an R-platform software tool, for our study. Data visualization, correctness, and statistical reliability are the main goals of the R-based Bibliometrix software. As open-source software, "R" has a broad supporting community that develops and integrates supplemental packages to enhance bibliometric analysis (Aria & Cuccurullo, 2017; Dervis, 2019). Bibliometrix integrates multiple statistical analysis modules at different levels and uses a workflow-based data analysis approach, unlike other free and open-source bibliometric analysis software like Vos Viewer and Cite Space.

The quantitative metrics in bibliometric analysis serve to assess the research output and various research networks objectively. This also implies that the research impact assessment can be evaluated while peer review is subjective. The methodology employed in this study is characterized by its transparency, replication of results, cost-effectiveness, relatively short time, and multi-level analysis—individual, institutional, national, and international levels. Bibliometric analysis has quantitative research output limitations, which can support the empirical research process to identify the research trends (Aria & Cuccurullo, 2017; Donthu et al., 2021; Zupic & Čater, 2015).

3 Findings and Discussion

Using Bibliometrix in R-package and Vos viewer, the bibliometric overview is displayed in descriptive and network analyses (Aria & Cuccurullo, 2017; Dervis, 2019; Donthu et al., 2021). The analysis is divided into two parts: performance and knowledge mapping. The descriptive analyses include information about the articles and citations per article and the ranking of the authors, papers, and journals. Through bibliographic coupling and term co-occurrence analysis, the network-centric approach explores the knowledge area intellectually and conceptually (Donthu et al., 2020, 2021; Kumar, 2015; Ninkov et al., 2022; Zupic & Čater, 2015).

3.1 Performance Mapping

In our corpus, 142 articles were published in 70 journals with an annual growth rate of 27.91%. These articles were written by 464 authors, of which 9 were written by single authors, with 4.04 co-authors per article with 32.39% of international co-authorship. The average number of citations received per document is 12.51, totalling 1777.

3.1.1 Annual Scientific Publications

The productivity and influence of publications are shown in Fig. 1, which is based on the total number of articles published and the total number of citations received.

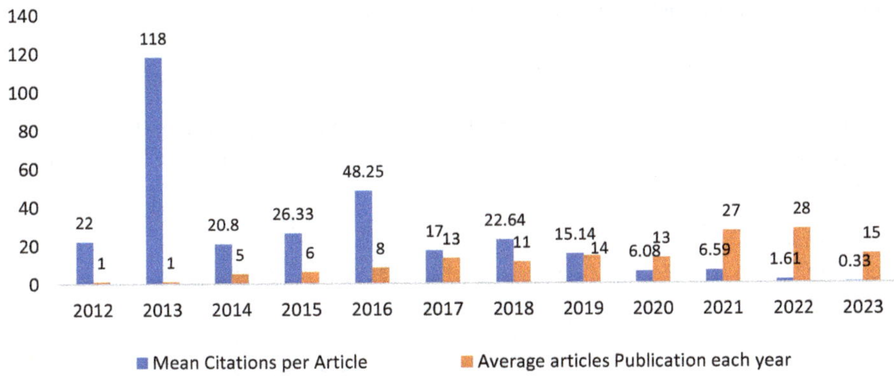

Fig. 1 Article citations and publication

Table 1 Top contributing authors and research papers

		Most cited articles and authors						
Authors	Articles	Paper	Total citations	Author	Year	freq	TC	TCpY
Anzer G	8	Rein R, 2016	229	Grossniklaus M	2018	1	92	15.33
Bauer P	8	Perin C, 2013	118	Janetzko H	2018	1	92	15.33
Janetzko H	6	Stein M, 2018	92	Keim Da	2018	1	92	15.33
Keim Da	6	Pappalardo L, 2019	59	Schreck T	2018	1	92	15.33
Schreck T	6	Stein M, 2017	52	Stein M	2018	1	92	15.33
Stein M	6	Andrzejewski M, 2015	51	Anzer G	2021	5	59	19.67
Brefeld U	5	Liu H, 2015	45	Bauer P	2021	5	59	19.67
Robertson S	5	Stein M, 2016	42	Grossniklaus M	2017	1	52	7.429
Woods Ct	5	Woods Ct, 2017	40	Janetzko H	2017	1	52	7.429
Grossniklaus M	4	Goes Fr, 2019	37	Keim Da	2017	1	52	7.429
Seebacher D	4	Yue Z, 2014	37	Schreck T	2017	1	52	7.429
Sinclair Wh	4	Stensland Hk, 2014	37	Stein M	2017	1	52	7.429

The number of published articles increased from 1 in 2012 to 28 in 2022 and 15 by mid-May 2023. Consequently, 2021 and 2022 show higher publications, with an increase in publications by 100% in 2021. The sports analytics phenomenon is growing in prominence in soccer sports. With 121 articles published since 2017, a significant increase is observed beginning in 2021.

Top Contributing Authors and Journals

Table 1 lists the top contributing authors (those with at least four articles). Anzer G. and Bauer P. have published eight articles, followed by Janetzko H., Keim Da,

Table 2 Yearly growth and number of articles published source wise

Journals' total publications and year-wise growth	Total Articles	2012	2013	2014	2015	2016	2017	2018	2019	2020	2021	2022	2023
International Journal of Sports Science and Coaching	31	0	0	2	7	9	12	13	15	20	23	24	31
Journal of Quantitative Analysis in Sports	6	0	0	1	1	1	2	2	3	3	4	6	6
Journal of Science and Medicine in Sport	6	0	0	0	0	0	0	1	3	4	5	6	6
Data Mining and Knowledge Discovery	5	0	0	0	0	0	2	2	2	2	3	5	5
Journal of Sports Sciences	5	0	0	0	0	0	1	1	2	2	3	5	5

Schreck T., and Stein M., with six publications each. Regarding the total number of citations, Grossniklaus M, Janetzko H, Keim Da, Schreck T, and Stein M were the top 5 authors with 92 citations each. Regarding papers, Rein R (2016) and Perin C (2013) received 229 and 118 citations, respectively.

The maximum contribution from the top five contributing journals that have published at least five articles is presented in Table 2. Here, we show the total number of research articles published in the journals and the year-wise growth in the publication. The *International Journal of Sports Science and Coaching* has published 31 research papers since 2014. The *Journal of Quantitative Analysis in Sports* and *Journal of Science and Medicine in Sport* published six research papers each. *Data Mining and Knowledge Discovery* and the *Journal of Sports Sciences* published five research papers each.

3.2 *Knowledge Mapping*

Knowledge mapping entails determining the underlying structure of the research domain under study that formed owing to conceptual and social interactions between the research sources. As "social structure", we analyse the country's scientific

Table 3 Countries' scientific production and most cited countries

Countries' scientific production		Most cited countries		
Region	Total articles	Country	Total citations	Average article citations
Germany	110	Germany	429	18.70
UK	66	Australia	203	12.70
Australia	64	USA	139	9.90
USA	55	France	118	118.00
Italy	32	Netherlands	89	29.70
Spain	26	UK	55	6.90
China	18	Spain	38	9.50
Norway	18	Norway	37	37.00
Netherlands	17	China	22	7.30
Brazil	15	Italy	22	4.40
Turkey	15	Poland	12	6.00
Austria	14	Turkey	12	4.00
Chile	12	Japan	11	11.00
Greece	11	Belgium	8	8.00
Poland	10			
Portugal	9			
Hungary	8			
South Korea	8			
Malaysia	7			
Iran	6			

production through author, institutions, and countries' collaboration, "conceptual structure" through co-word analysis (frequency and co-occurrence), thematic mapping, and thematic evolution through bibliographic coupling analysis (Donthu et al., 2021; Ninkov et al., 2022; Zupic & Čater, 2015).

Table 3 shows the countries with the highest publications in sports analytics and soccer research articles from 2012 to 2023 out of 37 countries. The research analysis shows that Germany leads with 110 research articles, followed by the UK with 66, Australia with 64 articles, the USA with 55 articles, and Italy with 32 articles, ranked second, third, fourth, and fifth, respectively. The analysis of 37 shows that 22 countries with less than 10 research articles, including Finland, Estonia, Denmark, Switzerland, Belgium, Argentina, and France, have less than 10 research publications. In these countries, sports analytics in soccer sports research has scope for exploration. Country publication data does not adequately reflect scientific contribution across nations and, therefore, needs to consider factors associated with top-performing countries. We are using the corresponding authors' details of 26 countries.

Table 4 shows the top 26 author countries and their single-country publication (SCP) and multiple-country publication (MCP) breakdowns. Corresponding authors belong to the same country affiliation in most SCP and MCP. Despite having the

Table 4 Corresponding author's countries

Corresponding author's countries					
Country	Articles	SCP	MCP	Freq	MCP_Ratio
Germany	23	15	8	0.162	0.348
Australia	16	10	6	0.113	0.375
USA	14	13	1	0.099	0.071
UK	8	4	4	0.056	0.5
Italy	5	5	0	0.035	0
Spain	4	2	2	0.028	0.5
China	3	2	1	0.021	0.333
Greece	3	2	1	0.021	0.333
Hungary	3	2	1	0.021	0.333
India	3	3	0	0.021	0
Netherlands	3	2	1	0.021	0.333
Portugal	3	2	1	0.021	0.333
Turkey	3	3	0	0.021	0
Iran	2	2	0	0.014	0
Korea	2	1	1	0.014	0.5
Malaysia	2	2	0	0.014	0
Poland	2	1	1	0.014	0.5
Argentina	1	0	1	0.007	1
Belgium	1	0	1	0.007	1
Brazil	1	1	0	0.007	0
Egypt	1	0	1	0.007	1
France	1	1	0	0.007	0
Israel	1	1	0	0.007	0
Japan	1	1	0	0.007	0
Norway	1	1	0	0.007	0
Serbia	1	1	0	0.007	0

most corresponding author affiliations, Germany, Australia, and the USA have MCP ratios (calculated by dividing the MCP by the number of articles published by each country) of 0.348, 0.375, and 0.071, respectively.

Table 5 shows the top 16 institutional affiliations having the most research output. The following are institutional affiliations, the University of Konstanz (published 19 articles), James Cook University (published 16 articles), Victoria University (published 13 articles), Deakin University (published 12 articles), and the University of Oslo (published 12 articles).

Word frequency analysis is a primary statistical method to determine research topic trends by identifying the most frequently appeared words in the research articles (Table 6). The most common words are thought to have more prominence in the research study and provide insights into key concepts of the current research domain. Co-occurrence is a semantic similarity in textual data in the co-word analysis. The keywords and authors' details are used for word frequency (word cloud) and

Table 5 Affiliations with the number of publications

Affiliations with the number of publications	
Affiliation	Articles
University of Konstanz	19
James Cook University	16
Victoria University	13
Deakin University	12
University of Oslo	12
University of Tübingen	10
Leuphana University of Lüneburg	8
Sheffield Hallam University	8
University of Groningen	8
Budapest University of Technology and Economics	7
Leeds Beckett University	7
Islamic Azad University	6
Politecnico Di Torino	6
Queensland University of Technology	6
Universiti Tun Hussein Onn Malaysia	6
Universtiät Mannheim	6

Table 6 Keyword analysis

Trend topics				
Topics	Frequency	Year_1st quarter	Year_medium	Year_3rd quarter
Computer graphics	8	2013	2016	2021
Spatial-temporal data	5	2016	2017	2020
Visual analytics	5	2017	2017	2018
Soccer player	6	2017	2018	2021
Visualization	6	2016	2018	2021
Multivariate analysis	5	2018	2018	2021
Humans	18	2019	2019	2022
Australia	10	2018	2019	2019
Data mining	8	2018	2019	2020
Football	39	2019	2020	2021
Soccer	28	2019	2020	2022
Human	22	2019	2020	2022
Sports	39	2017	2021	2022
Athletic performance	27	2019	2021	2022
Article	13	2019	2021	2022
Sport analytic	16	2022	2022	2022
Deep learning	9	2022	2022	2023
Forecasting	6	2021	2022	2022

Fig. 2 Word cloud of sports analytics and soccer research

co-citation analysis. Identify the keywords with similar meanings and clubbed words before executing the word frequency code (Brzezinski, 2015; Cuccurullo et al., 2016; Kumar, 2015). The word frequency is proportional to the font size in the word cloud. The following keywords appeared more frequently: football, sports, soccer, athletic performance, humans, ports analytic, competitive behaviour, deep learning, machine learning, computer graphics, and data mining (Fig. 2).

The co-authorship network shows the collaboration between the authors (Donthu et al., 2021; Kumar, 2015; Zupic & Čater, 2015). Figure 3 shows that most authors work in isolation as majority of the 464 authors are not connected, and the most extensive connected set of items is 42.

Country collaboration network explored collaboration on sports analytics and soccer. The sports analytics and soccer research has 17 clusters with 29 countries involved. The key three clusters dominate the international sports analytics and soccer research partnership network. Germany has the highest number of 27 collaborations, followed by the USA and the UK with 12 each (Fig. 4). We conclude that international cooperation in sports analytics and soccer research is slowly developing. However, intercontinental collaborations can improve and mature their partnership. Most of the research collaborations are associated with continental collaborations. Strong collaborations such as Germany, the USA, the UK, and Australia can support the cross-cultural and geographical interchange of ideas to advance the research area.

Co-occurrence of Words

The occurrence of word frequency kept at 2 words, which gave 144 keywords and formed 8 clusters (Table 7) and with 3 words, it gave 94 keywords and 6 clusters (Fig. 5).

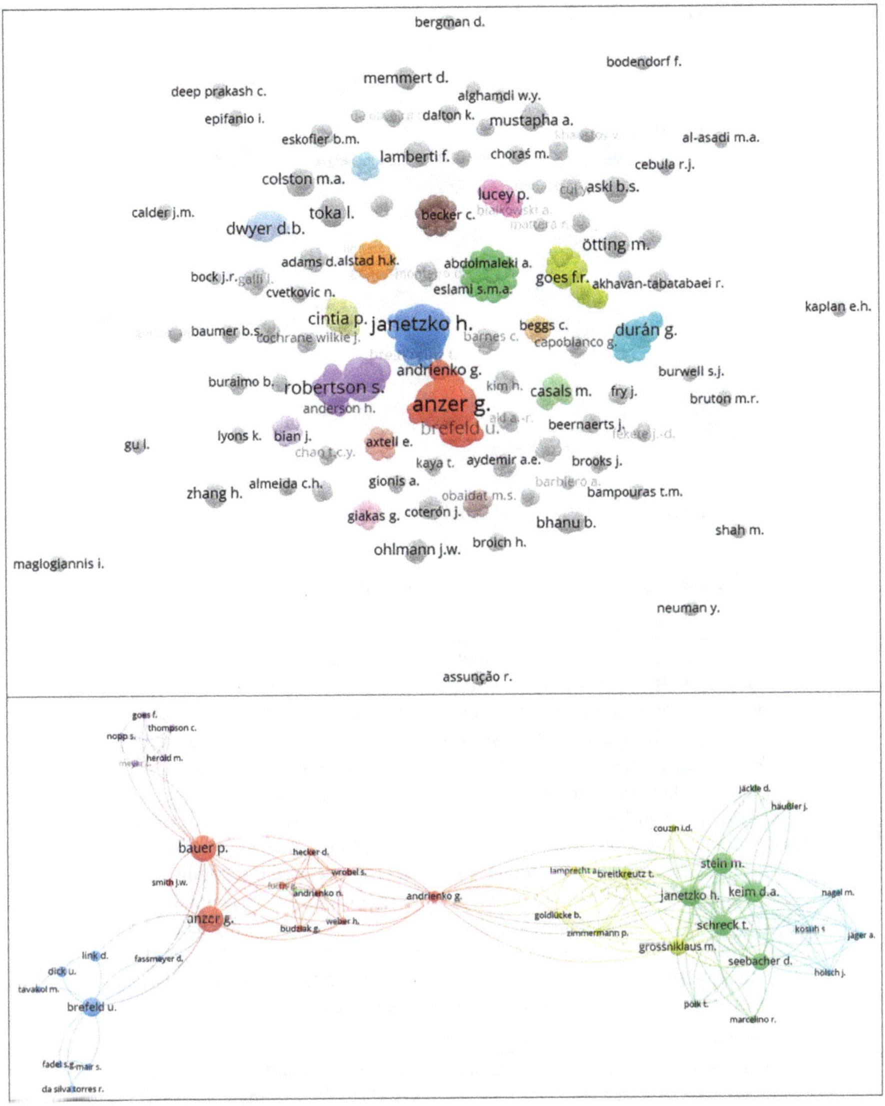

Fig. 3 Co-authorship network

Thematic Cluster Analysis

Thematic cluster analysis was performed to explore sports analytics in soccer sports conceptual structure. In which frequent keywords are interconnected (rule of association), have similar themes, and cluster together to indicate the research themes or concepts, such association may link keywords in separate clusters. Interrelated clusters or themes affect the research area. Each research theme produced through such

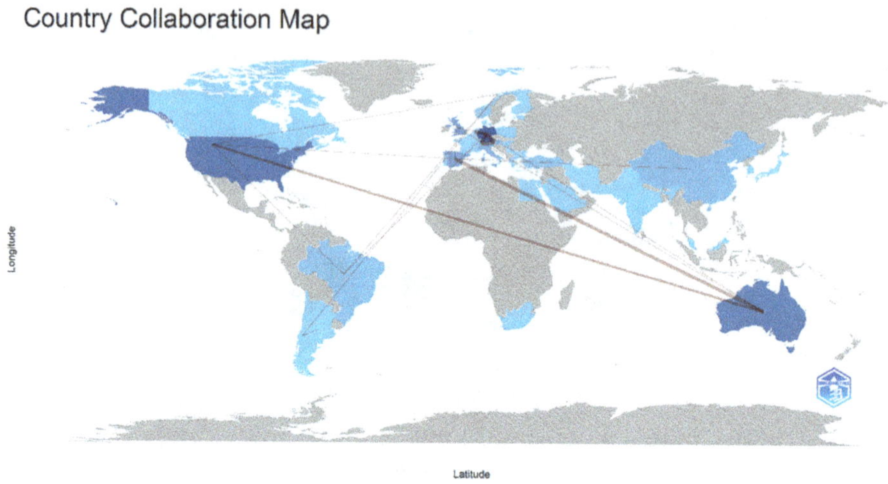

Fig. 4 Country collaboration map for sports analytics and soccer

Table 7 Clusters of co-occurrence of words of sports analytics in soccer

Clusters	Keywords (minimum 2 frequency)
One	Sport analytic, deep learning, computer graphics, data visualization, forecasting, learning systems, soccer player, visualization, feature extraction, match analysis, spatio-temporal data, visual analytics, cameras, data handling, decision making, learning algorithms, multi agent systems, performances analysis, predictive analytics, tracking data, computer vision, data acquisition, features extraction, football players, graph neural networks, human computer interaction, iterative methods, machine-learning, problem solving, sports data, state of the art, supervised learning, team performance, tracking (position), alignment, artificial intelligence, benchmarking, clusterings, computer vision techniques, data science, event recognition, game, gradient boosting, information analysis, internet of things, large dataset, long short-term memory, machine learning models, machine learning techniques, mean square error, movement analysis, multi agent, network architecture, pattern recognition, predictive models, procedures, soccer analytic, soccer analytics, supervised machine learning, team sports, trajectories, video analysis
Two	Classification (of information), convolution
Three	Risk assessment, athletic injuries, risk factors, cohort analysis, injuries, numerical model, performance assessment, physical conditioning, human, physical performance, physiology, receiver operating characteristic, sports medicine, universities
Forth	Mathematical programming, optimization
Fifth	Digital storage, performance analysis, predictive modelling
Sixth	Data mining, machine learning, multivariate analysis, big data, cluster analysis, decision tree or trees, longitudinal studies, retrospective studies, social media, team sport, movement or movement (physiology), prediction, principal component analysis, seasons, dependent variable, global positioning system, hierarchical clustering, major clinical study, theoretical model (models, theoretical), quantitative analysis, task performance, task performance and analysis, turnover rate, velocity
Seventh	Soccer video, a blob-based method, detection and tracking, a feature-based method, image processing, and occlusion detection
Eighth	Soccer analytics and sports analytics

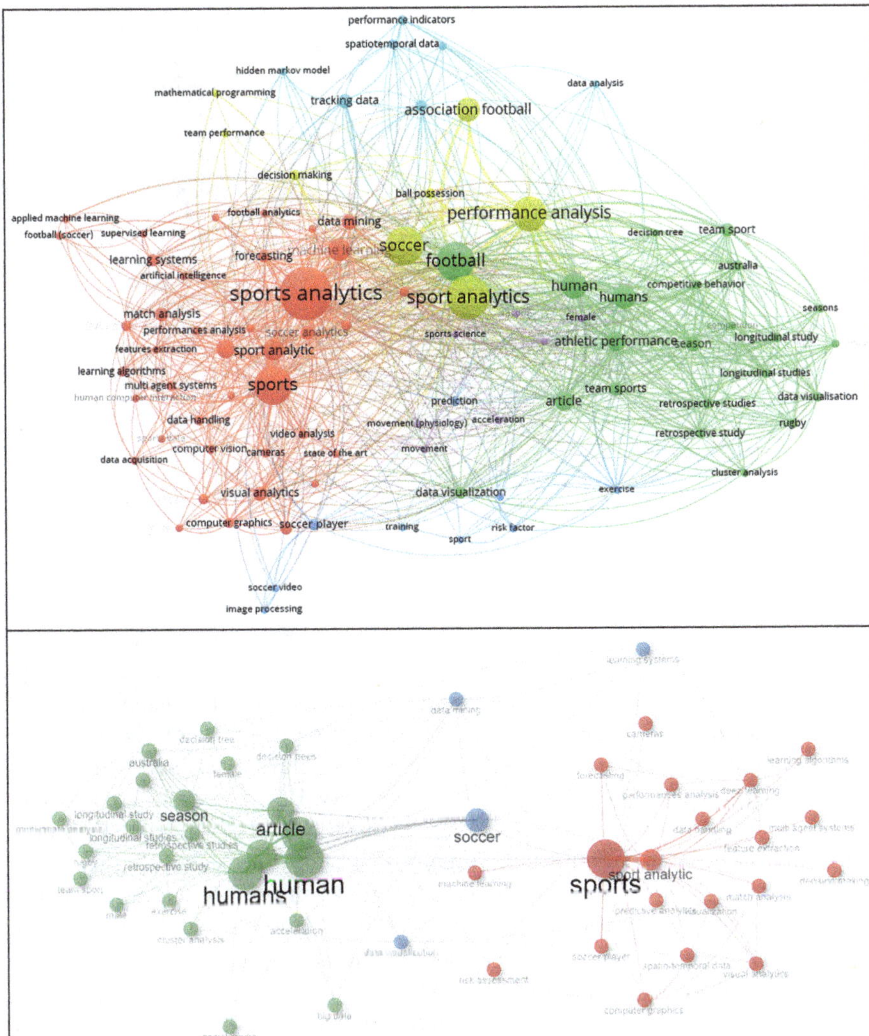

Fig. 5 Network of occurrence of word clusters of sports analytics in soccer

a clustering procedure has two parameters: "density", which determines the extent of relationships (appearing in combination in certain documents) among its constituent keywords, and "centrality", which assesses the degree of external ties (constituent keywords coming from one theme that occurs together with keywords from another theme). Density indicates the development of the theme, and centrality indicates its importance to the research field (Cobo et al., 2011; Donthu et al., 2020, 2021; Ninkov et al., 2022).

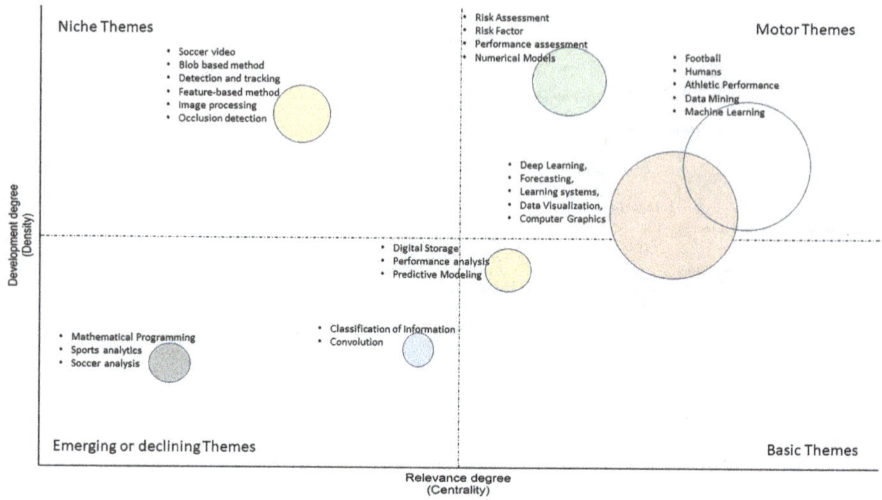

Fig. 6 Thematic cluster analysis of sports analytics in soccer

Thematic cluster analysis gave the most relevant themes, and we identified strategically essential themes presented in the diagram to analyse the research developments in sports analytics in soccer sports (Fig. 6). Thematic cluster analysis diagram is a two-dimensional representation created by plotting the research themes based on the development and connectedness on the x-axis (relevance degree—centrality) and y-axis (development degree, density), respectively. Based on centrality and density, research themes can be divided into four quadrants: motor themes with high centrality-high density, basic themes with high centrality-low density, highly developed and isolated themes with low centrality-high density, and emerging or declining themes with low centrality-low density. The theme cluster shows that the emergent themes (motor) reveal that methodological approaches such as machine learning (deep learning, forecasting), data visualization, and computer graphics are growing. At the same time, application areas like football, humans, and athletic performance are relatively developed, i.e. they have strong internal connections. This quadrant also shows the areas of concern like risk assessment, risk factors, assessments, and numerical models. The second quadrant in the upper left corner of Fig. 6 contains subjects such as soccer video, blob-based method, detection and tracking, feature-based method, image processing, and occlusion detection. These themes have low centrality but high density, reflecting their stronger internal ties but weaker external connections to other themes, representing these themes as highly developed and isolated themes with high internal development but less influence on sports analytics in soccer research. The third quadrant in the lower right corner of Fig. 6 contains subjects such as digital storage, performance analysis, predictive modelling, and machine learning emerging approaches at the intersection of the two quadrants. These themes have high centrality but a low density, which shows their stronger conceptual connections with other themes of the research domain. However, these themes require more development to achieve higher density.

The final quadrant in the lower left corner of Fig. 6 contains two theme clusters. The one cluster of topics, such as classification of information and convolution, have relatively high centrality and density compared to the second cluster, which contains the following areas—mathematical programming, sports analytics, and soccer analysis (Bauer & Anzer, 2021; Bonomo et al., 2014). The themes present in this quadrant are emergent or declining due to limited internal development and limited external connections with other themes. Based on the analysis of research, it shows that these themes are emergent with new developments in technologies and new methodological tools. These themes also indicated that these are emergent for some countries while research concentrated on a few countries. We believe that topics such as mathematical programming and classification of information and convolution neural networks provide novel approaches for optimization to sports analytics in soccer games (Alarcón et al., 2017; Bonomo et al., 2014; Dalton-Barron et al., 2022; Dobreff et al., 2022; Durán et al., 2022; Malamatinos et al., 2022).

4 Conclusion

This research presents a comprehensive bibliometric analysis of sports analytics in soccer and provides insights into existing and future research. The bibliometric analysis examines the characteristics of scholarly publications, such as citation patterns and author collaborations, to determine football sports analytics' core themes, trends, and influence. Research shows that sports analytics in soccer have grown since 2012. Sports analytics promotes data-driven approaches to football performance. Most researchers and practitioners use analytics methods such as advanced statistical methods, machine learning algorithms, mathematical computations, and big data analytics to discover trends, make informed decisions, and improve team performance. Sports analytics for soccer has multiple sub-areas, including player performance analysis, match analysis, injury prevention, analytics technical tools, and methodological developments. The collaboration networks of scholars, institutions, and countries advance soccer sports analytics. These collaborations can improve research and enable new football data analysis methods and technologies. There is an apparent isolation between developed and developing countries research on this topic. This bibliometric analysis highlights notable authors, institutions, and publications, which can help researchers and countries to explore this area. Sports analytics can help researchers, practitioners, and policymakers to improve individual and team performance and decision-making.

Future Research Scope and Limitations

The use of sports analytics in soccer has evolved in the last decade. This research shows that existing research can be expanded to address important future research directions by the scholars. What are the advantages of the existing sports analytics approaches for developing countries? How can opportunities offered by sports analytics be seized in the future? Existing research mainly pays attention to developed countries. To address the challenges of developing countries using sports analytics can be meaningful research. This research provides a comprehensive overview of various factors associated with sports analytics in soccer. A comparative analysis of football leagues, regions, and teams would help to explore the research focus, techniques, and trends, which help to identify the best practices and opportunities for collaboration and knowledge exchange. Sports analytics involves multidisciplinary perspectives from computer science, statistics, sports science, and psychology. This provides future research scope on developing this field using knowledge and idea exchange from various disciplines. Future studies can compare the bibliometric findings from different databases or do a combined analysis. Along with bibliometric analysis, systematic literature review and meta-analysis may help to explore this dynamic area.

This research has some limitations. We included articles from Scopus, a popular indexing platform for interdisciplinary research on engineering and social sciences that features publications on these emerging topics. Existing results can be further enriched by considering additional indexing databases, such as the Web of Science.

References

Ács, B., Kovács, R., & Toka, L. (2023). A career handbook for professional soccer players. *International Journal of Sports Science & Coaching*, 174795412311555. https://doi.org/10.1177/17479541231155598

Alarcón, F., Durán, G., Guajardo, M., Miranda, J., Muñoz, H., Ramírez, L., Ramírez, M., Sauré, D., Siebert, M., Souyris, S., Weintraub, A., Wolf-Yadlin, R., & Zamorano, G. (2017). Operations research transforms the scheduling of Chilean soccer leagues and South American world cup qualifiers. *Interfaces, 47*(1), 52–69. https://doi.org/10.1287/inte.2016.0861

Anzer, G., & Bauer, P. (2022). Expected passes: Determining the difficulty of a pass in football (soccer) using spatio-temporal data. *Data Mining and Knowledge Discovery, 36*(1), 295–317. https://doi.org/10.1007/s10618-021-00810-3

Aria, M., & Cuccurullo, C. (2017). bibliometrix: An R-tool for comprehensive science mapping analysis. *Journal of Informetrics, 11*(4), 959–975. https://doi.org/10.1016/j.joi.2017.08.007

Ariyaratne, M. K. A., & Silva, R. M. (2022). Meta-heuristics meet sports: A systematic review from the viewpoint of nature inspired algorithms. *International Journal of Computer Science in Sport, 21*(1), 49–92. https://doi.org/10.2478/ijcss-2022-0003

Bauer, P., & Anzer, G. (2021). Data-driven detection of counter pressing in professional football: A supervised machine learning task based on synchronized positional and event data with expert-based feature extraction. In *Data mining and knowledge discovery* (Vol. 35(5), pp. 2009–2049). https://doi.org/10.1007/s10618-021-00763-7

Bock, J. R. (2017). Empirical prediction of turnovers in NFL football. *Sports, 5*(1), 1–11. https://doi.org/10.3390/sports5010001

Bonomo, F., Durán, G., & Marenco, J. (2014). Mathematical programming as a tool for virtual soccer coaches: A case study of a fantasy sport game. *International Transactions in Operational Research, 21*(3), 399–414. https://doi.org/10.1111/itor.12068

Borghesi, M., Costa, L. D., Morra, L., & Lamberti, F. (2023). Using Temporal Convolutional Networks to estimate ball possession in Soccer games. *Expert Systems with Applications, 223*(March 2022), 119780. https://doi.org/10.1016/j.eswa.2023.119780

Brzezinski, M. (2015). Power laws in citation distributions: Evidence from Scopus. *Scientometrics, 103*(1), 213–228. https://doi.org/10.1007/s11192-014-1524-z

Cobo, M. J., López-Herrera, A. G., Herrera-Viedma, E., & Herrera, F. (2011). An approach for detecting, quantifying, and visualizing the evolution of a research field: A practical application to the Fuzzy Sets Theory field. *Journal of Informetrics, 5*(1), 146–166. https://doi.org/10.1016/j.joi.2010.10.002

Cuccurullo, C., Aria, M., & Sarto, F. (2016). Foundations and trends in performance management and public administration domains. *Scientometrics, 108*(2), 595–611. https://doi.org/10.1007/s11192-016-1948-8

Ćwiklinski, B., Giełczyk, A., & Choraś, M. (2021). Who will score? A machine learning approach to supporting football team building and transfers. *Entropy, 23*(1), 1–12. https://doi.org/10.3390/e23010090

Dalton-Barron, N., Palczewska, A., Weaving, D., Rennie, G., Beggs, C., Roe, G., & Jones, B. (2022). Clustering of match running and performance indicators to assess between- and within-playing position similarity in professional rugby league. *Journal of Sports Sciences, 40*(15), 1712–1721. https://doi.org/10.1080/02640414.2022.2100781

De Jong, L. M. S., Gastin, P. B., Bruce, L., & Dwyer, D. B. (2022). Teamwork and performance in professional women's football: A network-based analysis. *International Journal of Sports Science and Coaching*, 1–12. https://doi.org/10.1177/17479541221092355

Dervis, H. (2019). Bibliometric analysis using Bibliometrix an R Package using R Packages.pdf. *Journal of Scientometric Research, 8*(3), 156–160. https://doi.org/10.5530/JSCIRES.8.3.32

Dobreff, G., Molnar, M., & Toka, L. (2022). Optimizing and dimensioning a data intensive cloud application for soccer player tracking. *International Journal of Computer Science in Sport, 21*(1), 30–48. https://doi.org/10.2478/ijcss-2022-0004

Donthu, N., Kumar, S., & Pattnaik, D. (2020). Forty-five years of Journal of Business Research A bibliometric analysis. *Journal of Business Research, 109*, 1–14. https://doi.org/10.1016/j.jbusres.2019.10.039

Donthu, N., Kumar, S., Mukherjee, D., Pandey, N., & Lim, W. M. (2021). How to conduct a bibliometric analysis: An overview and guidelines. *Journal of Business Research, 133*(April), 285–296. https://doi.org/10.1016/j.jbusres.2021.04.070

Durán, G., Guajardo, M., & Zamorano, G. (2022). Mathematical models for rescheduling Ecuador's 2020 professional football league season disrupted by COVID-19. *Society and Economy, 44*(4), 360–377. https://doi.org/10.1556/204.2022.00013

Ellegaard, O., & Wallin, J. A. (2015). The bibliometric analysis of scholarly production: How great is the impact? *Scientometrics, 105*(3), 1809–1831. https://doi.org/10.1007/s11192-015-1645-z

Fernández, J., Bornn, L., & Cervone, D. (2021). A framework for the fine-grained evaluation of the instantaneous expected value of soccer possessions. In *Machine Learning* (Vol. 110(6)). Springer US. https://doi.org/10.1007/s10994-021-05989-6

Freitas, R., Volossovitch, A., & Almeida, C. H. (2021). Associations of situational and performance variables with defensive transitions outcomes in FIFA World Cup 2018. *International Journal of Sports Science and Coaching, 16*(1), 131–147. https://doi.org/10.1177/1747954120953666

Kumar, S. (2015). Co-authorship networks: A review of the literature. *Aslib Journal of Information Management, 67*(1), 55–73. https://doi.org/10.1108/AJIM-09-2014-0116

Maggon, M. (2022). A bibliometric analysis of Journal of Relationship Marketing (2002–2019). *Journal of Relationship Marketing, 21*(4), 324–351. https://doi.org/10.1080/15332667.2022.2080465

Malamatinos, M. C., Vrochidou, E., & Papakostas, G. A. (2022). On predicting Soccer outcomes in the Greek League using machine learning. *Computers, 11*(9). https://doi.org/10.3390/computers11090133

Matano, F., Richardson, L., Pospisil, T., Politsch, C. A., & Qin, J. (2023). Augmenting adjusted plus-minus in soccer with FIFA ratings. *Journal of Quantitative Analysis in Sports, 19*(1), 43–49. https://doi.org/10.1515/jqas-2021-0005

Mavrogiannis, P., & Maglogiannis, I. (2022). Amateur football analytics using computer vision. *Neural Computing and Applications, 34*(22), 19639–19654. https://doi.org/10.1007/s00521-022-07692-6

Ninkov, A., Frank, J. R., & Maggio, L. A. (2022). Bibliometrics: Methods for studying academic publishing. *Perspectives on Medical Education, 11*(3), 173–176. https://doi.org/10.1007/s40037-021-00695-4

Pappalardo, L., Cintia, P., Ferragina, P., Massucco, E., Pedreschi, D., & Giannotti, F. (2019). Player Rank: Data-driven performance evaluation and player ranking in soccer via a machine learning approach. *ACM Transactions on Intelligent Systems and Technology, 10*(5). https://doi.org/10.1145/3343172

Pritchard, A. (1969). Statistical bibliography or bibliometrics? *Journal of Documentation, 24*(4), 348–349.

Rahimian, P., Van Haaren, J., & Toka, L. (2023). Towards maximizing expected possession outcome in soccer. *International Journal of Sports Science and Coaching*. https://doi.org/10.1177/17479541231154494

Roumani, Y. F. (2022). Sports analytics in the NFL: Classifying the winner of the Super Bowl. *Annals of Operations Research, 325*(1), 715–730. https://doi.org/10.1007/s10479-022-05063-x

Seakins, D., Gastin, P. B., Jackson, K., Gloster, M., Brougham, A., & Carey, D. L. (2023). Discovery and characterisation of forward line formations at centre bounces in the Australian Football League. *Sensors, 23*(10). https://doi.org/10.3390/s23104891

Vinué, G., & Epifanio, I. (2017). Archetypoid analysis for sports analytics. In *Data mining and knowledge discovery* (Vol. 31(6)). Springer US. https://doi.org/10.1007/s10618-017-0514-1

Zupic, I., & Čater, T. (2015). Bibliometric methods in management and organization. *Organizational Research Methods, 18*(3), 429–472. https://doi.org/10.1177/1094428114562629

Application of Artificial Intelligence in Sports Analytics: Analysing the Ethical and Legal Perspectives

Ananya Pandey and Achyutananda Mishra

1 Introduction

Artificial intelligence (AI) played a role in the change of the world as we know it and expedited our capacity to function and grow. Recent times have witnessed the multidimensional use of artificial intelligence. Due to the rapid development of technologies, the use of artificial intelligence has become common now-a-days. It is being used in various fields like health, weather, finance, and art, including sports. According to research conducted in the year 2022, the estimated use of smart technology in sports shall rise to 20.3% globally between the years 2023 and 2030 (Hammes et al., 2022). AI in sports is novel to deal with. With the sports industry's continual development and expansion, the old paradigm of governance, i.e. the manual labour functioning in sports such as manual recording of time, calculating the player's strength and performance, etc., is being called into question necessitating the integration of smart intelligence, specifically the inclusion of AI. The advancement of AI and its organisational application are fast, but little is known about how these developments will affect the ethical and legal perspective in the field of its application (Ghosh et al., 2023). Use of AI in the field of sports is very promising having a transformative effect in the arena of sports. The way sporting competition evolves in modern society has been greatly impacted by the incorporation of analytics in sports-related AI. Artificial intelligence and statistical analytics are used in several sports to develop game strategy and increase fan engagement. Artificial intelligence is currently used in sports, from football to Formula 1, to

A. Pandey (✉)
Christ University, Bangalore, Karnataka, India
e-mail: ananya.pandey@res.christuniversity.in

A. Mishra
School of Law, Christ University, Bangalore, Karnataka, India
e-mail: achyuta.mishra@christuniversity.in

A. Mansurali et al. (eds.), *Sports Analytics*,
https://doi.org/10.1007/978-3-031-63573-1_10

instruct athletes and devise tactics, among other things. To put it another way, AI has a significant influence on how people read and watch sports. For a very long time, the sports sector has used analytics and statistics. Sports provide a rich setting for the use of artificial intelligence because they already assess any aspect that can be measured. The everyday effects of artificial intelligence are not limited to the world of sports. Together, AI and sensor technology can help gamers improve their skills. The use of AI in sports training is boosting the effectiveness of every sport for every participant by creating customised training plans for players and providing real-time feedback (Chmait et al., 2021). AI analysis of prediction can be used in sports to enhance wellness and productivity. Wearable technology users can learn about the stress sportsmen face, which helps to preserve their health. AI is able to recognise trends in game-playing techniques, methods, and vulnerabilities. The human species has enhanced its perspectives on a variety of subjects through the usage of AI systems, and intelligent systems have in turn helped us learn more about ourselves. Due to its endless potential, AI in sports has given a chance to reduce distraction and inaccuracy while making crucial decisions and enabled better decision-making (Gulen, 2022).

2 Brief Perspective on Sports Analytics

Traditionally the term analytics, which means the method used in science to identify and convey the significant trends that can be discovered in data through statistics, has passed down to the several common fields such as business, statistics, security, and many other domains (Jayal et al., 2018). Now it can be used in the realm of sports as well. Sports analytics is the phrase used to explain how athletes' behaviour and data are applied to improve their performance and tactics. Sports analytics is essentially the study of sports figures, including player positioning, corporate operations, and performance. Sports businesses, as well as individuals and teams participating in competitions, can profit from the data. The incorporation of quantitative and scientific rules into sports is commonly referred to as sports analytics (Morgulec et al., 2018).

3 Application of AI in Sports

The prospects for coaches, players, and teams are expanding greatly with the advent of AI and deep learning, i.e. a type of machine learning using artificial neural networks where input is processed through multiple layers to derive ever-more-complex features, in the sports industry. Some of the significant areas of application of AI in sports can be described as follows.

3.1 Monitoring the Athletes' Physical Condition During and After the Performance

The application of AI in the sports arena can be helpful in determining the condition of an athlete before and after the performance. An athlete's actions can be examined by AI algorithms, which can also measure biomechanics and spot misalignments that could result in injury (Smerdov et al., 2022). AI can identify minute variations in joint angles, muscle activation patterns, and general movement mechanics by analysing video records or motion capture data. These algorithms can then alert athletes and their coaches to potential problems like imbalances, overuse, and improper movement patterns, enabling them to make educated modifications to avoid injury. By creating wearables that gather information on levels of pressure and wear, sportsmen can avoid serious injuries such as real-time monitoring of heart rate, body temperature, performance optimisation, and customised training plan. But that's only the start. Teams can make use of AI to develop strategies, tactics, and play to their strengths (Brefeld & Zimmermann, 2017). Player performance analysis has reached unprecedented heights, thanks to AI. Using data and graphics to get insights into their teams' strengths and weaknesses on any given day, coaches can modify their teams' tactics and strategies to capitalise on their opponents' deficiencies.

3.2 Real-Time Management for Technical and Advanced Tactics

AI is crucial in developing sophisticated automated supervision of the athletes through statistically intelligent analysis, training, and other methods. Finding videos of the winning strategies and visual assessments of each player are a few instances of the real-time management using AI. Players and coaches can use this to create more immersive experiences and counterstrategies.

This boosted the popularity of numerous sports and effectively marketed each individual athlete. The combination of live games and associated marketing benefits the sports sector greatly. For instance, numerous virtual games to advertise special teams are available on the market prior to the start of IPL matches. Starting businesses now have the chance to invest in mobile app development for their related businesses (Apostolou & Tjortjis, 2019).

3.3 Personalised Training Model

AI and data science combined with the mapping of customised learning systems can facilitate transparent personal development. The application of machine learning to sports has recently changed the industry and made it more efficient. The training and dietary habits for athletes have significantly improved throughout time. The use of artificial intelligence in sports, however, has the potential to further the tailoring of training and dietary regimens (Hammes et al., 2022).

3.4 Promote Scientific and Effective Technological Competition Strategies

By analysing numerous parameters like speed, spin, and placement through a variety of advanced scientific and technological tactics with the aid of AI and deep learning in sports, AI can assist in boosting sports performance. Additionally, it raises the level of competitiveness among the sportsmen.

3.5 Transformation, Upliftment, and Upgrading of Sports Industry

Digital transformation is fundamentally driven by AI. For instance, businesses may respond to sudden developments or new trends faster and more successfully by utilising AI-powered procedures. As a result, AI algorithms are able to analyse probable outcomes and potentially reduce future dangers because they can learn from previous events and information (Mileva, 2023). While reducing any potential dangers, AI aids in the optimisation of corporate operations. Trainers and coaches may now obtain crucial information, which enables them to decide more quickly and effectively. For instance, an AI solution can process and analyse massive amounts of data when it is integrated into your operations. AI solutions can learn from these interactions and instances with enough practice, improving its suggestions and predictions (Rosen, 2023).

3.6 Analysing and Detecting the Offences Under the Sports Analytics

Video and image analysis is used by the justice system and judicial sectors to obtain information about individuals, objects, and behaviours to support criminal investigations. But the analysis of image and video data is incredibly labour-intensive and necessitates a large investment in subject matter experts. Due to the enormous amount of information, the quick pace of changing technologies like smartphones and operating systems, and the dearth of specialised staff with the knowledge to handle such information, video and image analysis is also vulnerable to human error (Croft, 2022).

The use of AI in the domain of sports aids in determining and establishing that the conduct of the athlete was outside the reasonable expectation and constitutes an offence. The offence is determined by analysing the type of sport, the level at which sport was conducted, nature of the sport, degree of force used, the amount of risk of injury, and state of mind of the defendant. One of the most essential offences to be taken into account in the field of sports is doping. The use of smart intelligence and technology in sports helps in identifying the offence of doping, the level of doping, and the substance consumed in doping. AI is helpful in determining the effect of the substance consumed during doping in the body and during the performance of sports activity.

AI-based technologies enable us to perform expert-level jobs and rectify human errors. Conventional software methods for facial identification and pattern analysis are limited to specific parameters like eye colour, eye shape, and eye distance. Beyond human comprehension, AI video and image algorithms learn difficult tasks and independently generate and determine complex facial recognition elements and parameters. These algorithms may be used to match faces, recognise guns and other items, and recognise complicated events like accidents and crimes, whether they are occurring in real time or after the fact (Rigano, 2018). In response to the needs of the criminal justice and law enforcement communities, the government has made investments in a number of areas, including the quality, speed, and specificity of data collecting, imaging, and analysis as well as the quality of contextual information.

3.7 Scouting and Recruitment

Today's recruiting teams are going above and above by incorporating AI into their scouting and hiring tool-kit. Teams gather gigabytes of data as they keep track of every action that takes place on the playing field, including swings in baseball, runs in football, and blocks in basketball. In fact, computer vision is utilised in sports to monitor players' body position and movement during games (McCabe & Trevathan, 2018). With the knowledge this data provides, teams may improve their hiring practices moving ahead, ensuring they create a successful team and acquire undervalued talents at a discount who can then be developed and sold for a profit in the future.

3.8 Match Predictions

Individuals have been attempting to shift a tonne of data in an effort to forecast the results of upcoming games and earn big money for years. In an effort to forecast sports results, researchers have looked at first- and second-serve percentages in tennis, the quantity of aces, backhand winners, and other factors. In the end, though, a person cannot accurately anticipate enough games to make a million dollars, nor can they analyse as much data as an AI-driven football algorithm prediction can

(Ramkumar et al., 2022). AI is also unable to correctly estimate every single match's result. But a forecasting algorithm can come far closer than a human can.

3.9 Sports Journalism

Sports journalism is a lucrative industry. Some could even argue that there is too much information to cover, particularly when it comes to statistics and data from the tournaments and minor leagues that happen virtually every day throughout the world (Boyle, 2017). AI is now easing such onerous tasks in other industries, and this also applies to sports journalism. Then, there is the little issue of the match reports. There are hundreds, sometimes thousands, of football matches played every day. AI bots are currently being used to produce precise match reports that explain the crucial incidents, provide statistics and data, and keep the reader interested to meet the demands (Ratiu et al., 2010).

4 Advantages of Uses of AI in Sports

Artificial intelligence in sports has already started to transform the field and elevate the sport to unprecedented levels. Even though statistics and quantitative analysis have long been crucial to comprehending sports, the development of artificial intelligence (AI) raises the possibility that these elements of the game, along with how it is played and how spectators are involved, may alter. The usage of artificial intelligence and technology in general is growing in the sports industry. It has played a significant role in the development and growth of sports both inside and outside of stadiums and has made it possible for all teams and players to perform at their best (Johnson, 2023).

Some of the significant advantages of the usage of AI in sports can be classified as follows (Peterson, 2023):

- Performance Analysis

AI can record sports motions with extreme precision by using sensors and cameras, allowing instructors to spot even the smallest deviations in form or technique. An important role here is played by the field of computer vision, which enables computers to understand visual data. Ultimately, this high-precision data helps athletes perform better by enabling them to execute motions consistently and effectively.

- Organised Training Plans

One of AI's most potent benefits is its ability to customise training regimens. AI is able to identify specific shortcomings in an athlete's performance and develop training programmes that specifically target such weaknesses. Each athlete's potential and performance may be maximised with this data-driven, customised training, which frequently outperforms conventional approaches.

- Injury Prevention

AI algorithms can forecast the chance of an injury based on an athlete's movement patterns, heart rate, and other physiological data by evaluating data from wearable devices. This lowers the chance of overuse injuries by enabling trainers and coaches to modify training schedules properly.

- Game Strategy

After processing hours of game footage in real time, sophisticated AI algorithms identify trends, player actions, and weaknesses in the team and their opponents. This gives coaches the ability to act quickly on tactical decisions based on AI recommendations, such as changing formations or making important substitutes.

- Player Development

Artificial intelligence (AI) tools are used by the sports industry to evaluate the actions of individual players. These tools measure things like ball possession, pass accuracy, and tackle efficiency to identify players who are performing exceptionally well or who are falling short during a particular match.

- Fan Engagement

Deeper insights help fans understand the sport on a different level, which increases their degree of engagement. Data enhances the viewing experience, and immersive viewing experiences can be made possible by technologies like computer vision, which help us comprehend what we see in a video stream and where the items are. Real-time stat overlays help to describe player movements and activities.

- Referee Assistance

Umpire assistance, which entails employing technology to help umpires make accurate choices during matches, is one application of AI in sports analytics. Ball tracking systems, which track the movement and trajectory of the ball in real time, are among the best illustrations of this.

- Performance Tracking

Wearable devices and sensors equipped with AI can track player movements, heart rate, and other vital statistics during games and practice. This data can be used for performance assessment and improvement.

- Athlete Recruitment

AI systems may evaluate player data to improve team dynamics, individualise training plans, and assess physical performance, health measures, and nutrition.

- Data-Driven Insights

AI is an effective tool to analyse the trends that are beneficial to make informed decisions about the players' signing patterns, team and working strategies, and related business operations.
Overall, AI's integration into sports has the potential to enhance performance, safety, and the overall fan experience, making it a valuable tool for athletes, coaches, teams, and sports organisations.

5 International Recognition of Sports Law

The legal recognition of sports is not universal throughout the world. Instead, different regions follow their different set of formulated rules and regulations. This is because there exists no jurisprudence so far in this area. Gradually the domain is gaining momentum and recognition in the name of sports law. Several countries have adopted the use of AI in the field of sports, thereby increasing the efficiency (Deeks, 2020). The regions and countries like EU, the USA, and China have adopted the significant use of AI in the field of sports.

5.1 European Union (EU)

The EU places a very little emphasis on regulating the role of AI in sports. A need for the discrete body to govern and deal with the matter of ethics and legal issues is required to ensure transparency, efficiency, and efficacy in the field of sports. However, Article 3 of the EU Treaty (WeForum, 2023) mentions the peace, well-being of people, healthy environment, etc. It discards the inclusion of sports as its place in the common market. It states that there is no room for sports functioning in the common market and how is it related to sports? The EU has a twin notion of European sports law and policymaking, which arose as a result of sports' exclusion from the purview of Article 3. According to several supporters of the theory of jurisprudence, it has been laid down that every facet of the law that applies to sports is the same as it is established in every social or legal category. Further, the critics of sports law analyse that cases involving sports are grounded as same as in diverse fields of law such tort or contract. The Lisbon Treaty of EU reveals a gulf between the constitutional principle, where it seems to carry great weight and law and policymaking in practice (Parrish, 2003). If we consider the application of AI in sports, just a few areas have been investigated thus far, and it is still regarded as being in its earliest stages. With the application of AI in sports, it becomes essential to adhere to the principles, rules and regulations of the General Data Protection Regulation (GDPR), which is an essential component of EU's data protection and privacy laws.

As per Article 22 of the GDPR (General Data Protection Regulation, 2018), the problems of accuracy, prevention and detection of data bias, consent of interested parties, data minimisation for the purpose of data processing, profiling, and automated decision-making have to be considered while applying AI. The EU lacks the base for the emergence of a policy/law in the form of a treaty/convention. Currently, the EU incorporated a sports policy which had its base as it took a dogmatic intent in sports, with the EU having a political interest in sports.

The EU General Data Protection Regulation (GDPR) will likely have an impact on AI organisations that meet the criteria for EU membership. In accordance with Article 22 of the GDPR, an 'information object shall not, except in exceptional circumstances, be the subject of an assessment solely based on automated procedures, particularly establishing a profile that generates legal implications concerning or which similarly materially affects it'. One allowed requirement is based on the data subject's voluntary and informed consent (Xian, 2010). This will probably have an impact on how businesses address bias and transparency in AI.

5.2 *United States of America*

In the USA, contract law, intellectual property law, labour law, and tort law frequently intersect with sports law. The fundamental concerns and questions in the area of sports law relate to issues like the invasion of data privacy and worker rights. The legal landscape for AI in the USA is evolving as lawmakers and regulators grapple with the challenges posed by this rapidly advancing technology. Legal frameworks and regulations are expected to grow to address AI's unique legal and ethical issues. Organisations that develop, use, or deploy AI should stay informed about these developments to ensure compliance and mitigate legal risks. Artificial intelligence systems routinely process large amounts of personal data. Laws like the California Consumer Privacy Act (CCPA) and the federal Children's Online Privacy Protection Act (COPPA) control the collection, use, and sharing of personal data by AI apps. Liability for judgements or actions made by AI might be difficult to determine. In cases where AI causes harm or makes judgements that have substantial repercussions, courts may need to develop liability frameworks. AI use in sports carries a significant risk of violating workers' rights. There is currently no comprehensive federal law governing the use of artificial intelligence, no substantial state law (although state privacy laws may apply to AI systems that process certain types of personal data), and no serious discussion of a US version of the EU AI Act (Greenstein, 2022).

5.3 China

Since 2021, government officials in China have released new regulations on a monthly basis. These regulations include the Personal Information Protection Law, a national privacy law, and the Code of Ethics for New-Generation Artificial Intelligence, which outlines standards for new artificial intelligence (AI) solutions, such as 'improving human well-being' and 'controllability and credibility' (Heath, 2023). Since 2022, Chinese users of AI have been granted transparency rights, which include the capacity to turn off algorithmic advice services and the awareness of when they are exposed to content created by AI. Chinese AI companies must censor material and pass security checks in order to maintain Communist Party control prior to the release of new products, which adds to the difficulty and burden of China's journey to AI trust. New rules on deep-synthesis technologies, such as deepfakes and other generative AI systems, were adopted by China and were published by the Cyberspace Administration. The legislation imposes stringent requirements for AI-generated media, such as the need for identifiers like watermarks (Henderson, 2020). This comes as a number of new tools have been released to identify AI-generated content, which may be used for a variety of purposes, such as data theft, disseminating misinformation, and more.

6 Ethical Challenges

The failure to operationalise data and AI ethics while applying AI in sports frequently puts organisations at risk for reputational, legal, functional, and regulatory problems in the sports industry. The essence of AI ethics is to continually inquire about the fast-imposing technologies on human life (Bostrom & Yudkowsky, 2018). AI systems are growing significantly in terms of the quantity of data they use and the processing power they require. In terms of the scope of their deployment and the degree of responsibility they take on, their pervasiveness in society surpasses that of computing in the PC and Internet eras (Dignum, 2018a). But as the technology grows larger, many elements become unintelligible to even the most experienced practitioners, especially in its deep learning version. Here are a few of the ethical issues with AI.

6.1 Automation Prejudice

A major ethical challenge in the creation and application of AI systems is automation prejudice, often known as algorithmic bias or discriminatory automation. It refers to the possibility for AI algorithms to deliver biassed or discriminatory results, frequently as a result of biassed programming or training data, which might unfairly harm particular groups of people. Psychology has long investigated human prejudice in great detail. It is the outcome of implicit associations, demonstrating how unintentional prejudice can affect the course of events. In recent years, society has started to question just how much these preconceptions from people may permeate AI systems, with potentially disastrous results (Zhang, 2020). Algorithmic bias in AI systems can take many forms; racial prejudice and gender bias are only two examples. A crucial ethical requirement for AI development is to address prejudice in automation. Developers and organisations can design AI systems that support justice, equity, and favourable societal outcomes by acknowledging the possibility of bias and taking proactive measures to reduce it. To ensure that AI technologies benefit everyone, regardless of their background or qualities, ethical considerations are essential (Schreck et al., 2023).

6.2 Security and Privacy

The conventional triad of confidentiality, integrity, and availability is used to evaluate security. The level of an AI's resilience determines how well it can withstand and recover from an attack. With increasing power comes the potential for both positive and bad uses of the technology. This law applies to robots aimed at supplanting human soldiers or autonomous weapons as well as AI technologies that have the potential to cause harm when used maliciously. Cybersecurity will become much more crucial because these battles won't just take place on the battlefield (Kaur, 2023). Adopting AI in an ethical manner entails being accountable, responsible, and, most crucially, transparent. It unquestionably involves adhering to legal requirements, social conventions, corporate beliefs, and customer expectations. In recent years, ethical AI has also included data protection, specifically not disclosing biassed results.

6.3 Behaviour Manipulation

One ethical worry with AI in observation that goes beyond data collection and attention-grabbing is the use of information to subvert autonomous rational decision-making in both online and offline contexts. While attempts to manipulate behaviour are not new, applying AI systems could offer them a new perspective. Users' frequent interaction with data systems and the wealth of personal information they provide makes them vulnerable to 'nudges', manipulation, and deceit. With enough prior information, algorithms can be used to target specific people or small groups with input that is most likely to have an impact on them (Madiega, 2019). While the USA and China favour expansion with less regulation, likely in the expectation that this will provide them a competitive edge, the EU General Data Protection Regulation (Regulation (EU) 2016/679) has enhanced privacy protection. It is obvious that government and corporate actors have improved their capacity to violate people's privacy and influence them with the aid of AI technology and will continue to do so unless restrained by legislation that serves the interests of society at large (Petropolouos, 2018).

6.4 Accountability

The need to defend and explain one's decisions and actions to other stakeholders, users, and system partners is known as accountability. To ensure accountability, decisions have to be based on and justified by the algorithms employed for decision-making. This involves the need for depiction of the moral standards and social norms that are common in the operational environment and that the participant uses to guide their decisions. Both the role of guiding action (by generating beliefs and making decisions) and the function of explanation (by putting decisions in a larger perspective and by categorising them along moral criteria) are necessary for accountability in AI (Dignum, 2018a, b).

6.5 Responsibility

In the development, implementation, and application of (AI) systems, responsibility is a crucial ethical factor. It includes different facets of decision-making, accountability, and transparency in relation to AI technologies. Responsibility is the act of taking ownership of one's actions and the ability of AI systems to identify errors and unexpected results and hold people responsible. As the application of AI grows, concrete mechanisms are required to connect the judgements made by AI systems with the ethical handling of data and the conduct of the parties involved (Kaur, 2023).

6.6 Transparency

Transparency is related to the requirement to govern the data required to create these systems and to analyse, duplicate, and explain the mechanisms by which AI systems make decisions and learn to adapt to their environment. The AI algorithms used today are essentially opaque. Regulators and users, however, expect justification and transparency regarding the data used. Methods are required to manage data, its provenance, and dynamics as well as to analyse algorithms and their output (Dignum, 2018a, b).

6.7 *Explainability*

Explainability is a crucial concern in the context of AI. People could find it difficult to understand how judgements are made since AI systems can be challenging to decode. To ensure that AI systems are unambiguous and understandable (Neznamov, 2019), the concerned authorities need to ensure that they are able to justify and exhibit the accuracy, fairness, and transparency of the decisions rendered by their AI systems.

The following few considerations should serve as guides when assessing the ethical concerns of using AI in sports analytics:

- Sense of proportion and non-harm
- Social welfare
- Equality
- Inclusivity and impartiality
- Fairness
- Security and preservation of privacy
- Security and dependability
- Clarity and comprehensibility
- Accountability
- Alignment's significance
- Governable
- Human-centred strategy
- Sustainability

The earlier ICT and computing advancements have been withdrawn by AI, and it will encounter many ethical issues. Technology may be used for good, but it also has the potential to be abused (Naik et al., 2022). The distinction between a computer and a human could be blurred if AI is overly anthropomorphised and humanised. As AI develops further, there will be a new 'digital divide', whereby certain geographic and socioeconomic groups would profit from technology at a faster rate than others. AI will also have incalculable consequences on our environment and ecology.

7 Legal and Regulatory Aspects of AI in Sports

Legal and regulatory frameworks are necessary in the rapidly developing field of artificial intelligence (AI) to guarantee that AI is produced and used in a responsible and ethical manner. Businesses working in the AI industry will need to be aware of these governance and legal problems and make sure they are abiding by all applicable rules and laws (Kemp, 2018). Legal and ethical concerns must evolve in tandem with the growing and changing applications of AI in order to ensure that the technology is created and applied in a responsible and ethical manner. Some of the legal challenges and aspects to be taken into account while considering the use of AI in the field of sports are the following.

7.1 International Administration

Given that AI transcends national bounds, international governance is an important legal issue. It raises concerns about rules governing artificial intelligence as well as global governance. To ensure that AI is managed consistently and uniformly across borders, laws and regulations will be required (Demidov et al., 2022). Companies need to make sure they understand and follow all applicable international laws and regulations regarding AI systems.

7.2 Liability

Liability is one of the legal concerns that is frequently raised when talking about AI. There is currently no mechanism in place to hold these machines accountable for their actions (Vakkuri et al., 2022). Few surveys resembled the liability concerns of who is accountable for the actions and inactions of a machine learning system, with 80% of respondents stating it was a barrier. In addition, according to 81% of respondents—of whom 46% said this was an urgent need—legislation is required to define culpability in the context of AI (Briz and Bender, 2021). According to the 'Civil liability regime for artificial intelligence report', which was commissioned by the European Parliament, liability laws have a significant impact on the economy and society and are crucial to the advancement of AI. Liability laws encourage risk mitigation and discourage participation in dangerous activities. Additionally, they offer legal certainty, which enables businesses to accurately assess risks and develop their goods appropriately, which promotes innovation. Certain factors to be taken into account while considering the factor of liability are as follows.

7.2.1 Autonomy, in Regard to the AI in Usage

In the context of AI, autonomy is the capacity of artificial intelligence systems to carry out operations autonomously and frequently without human supervision. There are significant ethical and pragmatic issues raised by the vast range of autonomy. Strict responsibility is a legal doctrine that states that the person in control of an artificial intelligence system may be held liable for any harm caused by the system's actions, regardless of guilt. Guidelines and rules are being developed by governments and regulatory agencies more frequently to handle liability issues associated with AI. Depending on the industry and the jurisdiction, these rules may change (Rodrigues, 2020).

7.2.2 Connectivity, with Other System Handling and Managing Data

A cohesive and efficient technology ecosystem requires connectivity between AI systems and other data-handling and management technologies. The seamless integration and feeding of data from multiple sources into AI systems is made possible via connectivity. This comprises unstructured data, streaming data from sensors, structured databases, and data from external APIs.

7.2.3 Opacity, Determining the Burden of Proof

Opacity in AI is the absence of transparency or comprehensibility in the decision-making processes of AI systems. Understanding how AI systems arrive at their conclusions or predictions can be difficult when they operate with a high level of opacity. The burden of proof in relation to responsibility in AI-related incidents can be significantly impacted by this opacity. The plaintiff frequently bears the burden of proving that the defendant's acts actually caused harm in traditional liability trials. To prove that the AI system did not hurt someone as a result of carelessness or design defects, however, may fall to the entity deploying or running the AI system in circumstances involving opaque AI systems. The software developer owes it to his or her client to uphold the required standards of care, failing which they risk legal action for a number of reasons, including:

- Failure of the developer to spot flaws in the features and functionalities of the programme
- A knowledge foundation that is insufficient or improper
- Improper or insufficient notices and documentation
- Failing to keep a current knowledge base
- Error brought on by incorrect user input
- Excessive user reliance on the result
- Misusing the software (Liao, 2020)

7.3 *AI and Legal Bias*

Two further important legal difficulties related to AI are bias and prejudice. AI systems have the power to legitimise bias and prejudice in decisions. To ensure that AI systems are unbiased and equitable and that individuals who have been the targets of discrimination have legal redress, laws and regulations will be necessary. Businesses and sports must make sure they are aware of any potential biases in the data they are utilising and put detection and correction procedures in place (Niziol, 2021).

7.4 AI and Right to Privacy

One of the biggest legal issues with AI is the protection, privacy, and security of data. Artificial intelligence (AI) systems employ vast amounts of data to learn and make predictions. In order to ensure that personal information is safeguarded and people's rights are upheld, regulations pertaining to data privacy and protection, such as the General Data Protection Regulation (GDPR) (Olivia, 2020), will be essential. In the context of sports, the designated authorities are required to take appropriate organisational and technical steps to safeguard personal information, obtain persons' informed consent prior to collecting and processing their data, and ensure that they are in accordance with all applicable data protection laws.

7.5 AI and Intellectual Property

AI development and use, along with allied technologies like computer hardware, might raise particular intellectual property (IP) problems with relation to possession and violation in the domain of sports. A distinct issue that needs to be considered in the context of AI and IPRs is the practical difficulty of patenting AI systems under the subject-matter eligibility condition. Unless they have been given a technical character in the form of useful software, algorithms by themselves typically qualify as a vague system lacking technical character and cannot therefore be protected under IP rules (IndiaAI, n.d.). However, IP regulators in several countries have also developed guidelines about examination and attempts to promote patent protection in these technological fields in response to the changes in technology and rising applications. They make it clear which algorithms are eligible to be patented so that it is possible to evaluate them later (Dayal, 2022).

The ownership, user, and authorship rights of information as well as ideas that AI systems generate on their own are important topics for discussion. At this point in time, however, there aren't many instances of content produced by completely autonomous AI systems; 'independent acting' computers are still a long way from being more commonplace in society. However, this sparks a discussion over whether the ideas and material produced with AI systems are patentable or copyrighted. The majority of states now forbid the issuance of such protection to an AI system because the current IPR laws only grant rights to entities accorded a legal personhood (whether natural or corporate) (Davies, 2011). However, since AI technology is still in its infancy and 'autonomy' and 'AI' have ambiguous definitions, legislating on this topic seems to be a challenge.

8 Need for a Robust Regulatory Framework for Sports Analytics

AI is a vital part of the technical operating system and is always growing and evolving on its own. As a result, laws controlling this area should either be universal to be effective regardless of advancements in information technology or be continuously revised to be effective regardless of advancements in information technology (Cerka et al., 2015).

An intriguing new era in sports analytics and player performance forecasting is underway, with AI driving the charge in predictive modelling. Several national and international sports federations and agencies use AI for improving the performance of players, predicting injury, experiencing the breakout session during the game, automated sports journalism, and much more. The field of sports analytics and AI is being regulated by several countries independently such as the EU, China, and the USA. A need for concrete, uniform, and global policy is highly needed in the contemporary world (Jha, 2020).

The development of accountable artificial intelligence is of paramount importance, and it can be accomplished by taking the following actions.

8.1 Policy Changes to Facilitate the Appropriate Adoption of AI

It's critical to control and raise the bar for ethical AI in India since it may have a big impact on athletes' functionality and performance. To assure the ethical application of AI in the world of sports, study into the technological, legal, policy, and social elements of responsible AI is required (Chandan, 2021). In order to ensure the responsible use of AI, it might also be beneficial to enable access to data and utilise responsible AI tools and techniques with some strict restrictions and means.

8.2 Responsible and Ethical AI Research

Making COREs (Consortium of Ethics Councils at Each Centres of Research Excellence) and ICTAI (International Centre for Transformational AI) determine the required practices and monitor their adoption would be of utmost importance and significance to ensure the implementation and compliance of ethical and legal aspect of the use of AI in sports (Olfers, 2009). Additionally, creating data trusts that will enable innovative, safe, and ethical uses of data, including the employment of AI in the field of sports, must be given careful consideration. Additionally, the development of a legal and regulatory framework for data privacy might be helpful for the sports industry in addressing and implementing the data protection framework, which could guarantee privacy and human rights without impeding innovation in India. Additionally, working with the various industries to develop sector-specific privacy, security, and ethics norms for communication, robotics, identity, sports, etc. could help to safeguard the set of data that is currently available. It is also imperative to work together and get funding from the COREs and other international organisations in order to preserve AI technology research. In order to fill the gap in the failure to strengthen data privacy, the adoption and implementation of new data protection and privacy legislation that adheres to the EU's GDPR framework is essential. Additionally, India's privacy protection laws would need to be regularly modified to account for newly developing hazards and their effects.

By completing damage impact assessments at each stage of the creation of the AI model, the participants can self-regulate and carry out the negligence test for damages brought on by AI software. Additionally, safe harbours for the analytics must be developed in order to protect against or limit the liability associated with designing, testing, monitoring, and improving the AI efficiency. For the parties to bear proportionate accountability for harm produced by products with integrated AI, a framework for apportioning damages can be engineered (Vakkuri and Abrahamsson, 2018).

8.3 Continuity of Research

The establishment of a Centre for Studies on Technology Sustainability (CSTS) is crucial in order to address the concerns regarding the ethical, legal, social, and international viability of the technologies developed. Research should focus on developing better data collection methods, such as advanced sensors, cameras, and wearables, to capture more detailed and comprehensive data from athletes and games. Additionally, improving data quality through noise reduction and error correction is essential. It will take research to create more effective techniques for combining and coordinating these disparate data sources in order to present a complete picture of individual and team performance. Enhancing AI models' ability to predict player and team performance is crucial. Researchers can work on developing more sophisticated predictive models that consider a broader range of factors, such as player psychology, team dynamics, and external variables, such as weather conditions. Long-term studies that track the impact of AI-driven insights on player development, team performance, and injury prevention can provide valuable feedback for continuous improvement and validation of AI algorithms. Continued research in these areas will help drive innovation and further enhance the role of AI in sports.

8.4 Developing Accountable Artificial Intelligence Competence and Understanding

In order to ensure that projections from AI are realistic and the enabling variables for the success of AI efforts are fully understood, the key problem is raising consciousness around the potential of AI. Identifying and encouraging the readily accessible datasets, legislative actions, and other tools is necessary to allow ethical AI in India. Other approaches to addressing the issue of data theft and data privacy breaches include reducing information asymmetry, addressing trust concerns and apprehensions about AI systems, and improving one's ability to recognise and analyse ethical dilemmas.

Without a question, artificial intelligence will keep changing sports, and these changes will bring with them brand-new and surprising ways for us to participate in, observe, and analyse sports. In reality, machine learning has fundamentally altered how we think about player performance metrics and match plans, as well as how we track, categorise, and study sport consumers (Duggal, 2017). A Pandora's box of ethical concerns will open up when machines are introduced into the traditionally human-centred and naturally gifted athlete foundation of sport; these questions will need to be considered more often. While it is unlikely that AI will completely replace human specialists and coaches, there is no doubt that coaches and players who properly use AI will have a major advantage over those who just rely on human expertise. Without a doubt, coaches and athletes who fully utilise AI technology will have a substantial advantage over those who only rely on human knowledge. But it's unlikely that AI will ever completely replace human specialists and instructors. Executives in the sport industry will also gain deeper, up-to-date insights on the inclinations, needs, and actions of sport fans. AI will therefore be crucial in producing sports material that is customised and suited to individual customers. However, it appears that, at least in the near future, human guidance and involvement will continue to be crucial for achieving top athletic performance and making strategic decisions in the sports industry (Novatchkov & Baca, 2013).

9 Conclusion and Suggestions

AI is still at that stage that demands improvement. The people associated with the field of sports, i.e. coaches and athletes, fail to realise the importance of AI in sports. Consequently, the development of sports intelligence and the creation, advancement, and execution of pertinent legislation pertaining to the usage of AI in sports is the need of an hour. A concrete rule governing the nature and application of AI in sports is necessary to guarantee athlete safety and sports fairness. In order to provide the privacy support to the athletes and overcome the technical loopholes and limitations, the clarification over the responsibility of the government has to be taken into account. Certain strategies such as strengthening the available policies in specific regions, enhancing financial support to the concerned governing authorities regulating AI, improving the standard of application of smart intelligence in the domain of sports, use of several intelligent products, upholding the fairness and safety of sports, establishing scientific and sound technological data information standards, and thereby improving the training of staff of the sports authorities may lead to utilisation of AI in graciousness. In contrast to the EU, there should be a General Data Protection Regulation (GDPR) rule that applies to all applications of AI in various industries around the world and addresses ethical and legal issues (Goddart, 2017). According to studies and experience, AI will inevitably replace entire divisions of labour, especially in professional fields, transportation, marketing, sports, and customer support. Data collection and the creation of an ethical risk framework for AI are useful tactics for reducing ethical concerns. It ought to consist of an upholding governmental structure. The issues of using AI in sports analytics are ethical and legal, and there are ethical councils that promise assistance, tools, and awareness as answers. Therefore, to make full use of AI's benefits and preserve the fairness and integrity of sporting events, it is imperative to strike a balance between innovation, ethical responsibility, and legal compliance.

References

Apostolou, K., & Tjortjis, C. (2019). Sports Analytics algorithms for performance prediction. In *10th International conference on information, intelligence, systems and applications (IISA)* (pp. 1–4). IEEE.

Bostrom, N., & Yudkowsky, E. (2018). The ethics of artificial intelligence. In *Artificial intelligence safety and security* (pp. 57–69). Chapman and Hall/CRC.

Boyle, R. (2017). Sports journalism: Changing journalism practice and digital media. *Digital Journalism, 5*(5), 493–495.

Brefeld, U., & Zimmermann, A. (2017). Guest editorial: Special issue on sports analytics. *Data Mining and Knowledge Discovery, 31*, 1577–1579.

Briz, N., & Bender, A. (2021). *Liability for AI decisions*. Dentons. https://www.businessgoing.digital/key-challenges-of-artificial-intelligence-liability-for-ai-decisions/

Cerka, P., Griginie, J., & Sirbikyte, G. (2015). Liability for damages caused by artificial intelligence. *Computer Law and Security Review, 31*, 376–389. https://doi.org/10.1016/j.clsr.2015.03.008

Chanda, M. S. (2021). *Handbook: Developing sports law in India: A challenge ahead*. INSC International Publishers.

Chmait, N., et al. (2021). Artificial intelligence and machine learning in sports research: An introduction for non-data scientists. *Frontiers in Sports and Active Learning, 3*, 682287. https://www.frontiersin.org/articles/10.3389/fspor.2021.682287/full

Croft, H. (2022). *Addressing the problem of 'big data'in sports: A framework for performance analysts* (Doctoral dissertation, Auckland University of Technology).

Davies, C. R. (2011). An evolutionary step in intellectual property rights–Artificial intelligence and intellectual property. *Computer Law & Security Review, 27*(6), 601–619.

Dayal, H. (2022). AI and IPR: Future challenges. *International Journal of Corporate Law and Policy*. https://ijclp.com/artificial-intelligence-ai-and-intellectual-property-rights-ipr-legal-status-and-the-future/

Deeks, A. (2020). Introduction to the symposium: How will artificial intelligence affect international law? *AJIL Unbound, 114*, 138–140. https://doi.org/10.1017/aju.2020.29

Demidov, V., Dolzhenkova, E., Mokhorov, D., & Mokhorova, A. (2022). Legal aspects of artificial intelligence application in artistic activity. In *Technology, innovation and creativity in digital society: XXI professional culture of the specialist of the future* (pp. 192–202). Springer International Publishing.

Dignum, V. (2018a). Ethics in artificial intelligence: Introduction to the special issue. *Ethics and Information Technology, 20*(1), 1–3.

Dignum, V. (2018b). *The art of AI*. The Medium. https://medium.com/@virginiadignum/the-art-of-ai-accountability-responsibility-transparency-48666ec92ea5

Duggal, P. (2017). *Artificial intelligence law*. Independently Published.

General Data Protection Regulation. (2018, May). https://gdpr-info.eu/

Ghosh, I., Ramasamy Ramamurthy, S., Chakma, A., & Roy, N. (2023). Sports analytics review: Artificial intelligence applications, emerging technologies, and algorithmic perspective. *Wiley Interdisciplinary Reviews: Data Mining and Knowledge Discovery, 13*, e1496.

Goddard, M. (2017). The EU General Data Protection Regulation (GDPR): European regulation that has a global impact. *International Journal of Market Research, 59*(6), 703–705.

Greenstein, S. (2022). Preserving the rule of law in the era of artificial intelligence (AI). *Artificial Intelligence and Law, 30*(3), 291–323.

Gulen, K. (2022). *Does AI spoil the naturalness of sports?* Data Economy. https://dataconomy.com/2022/11/03/artificial-intelligence-in-sports-examples/

Hammes, F., Hagg, A., Astaroth, A., & Link, D. (2022). Artificial intelligence in elite sports—A narrative review of success stories and challenges. *Frontiers in Sports and Active Living, 4*, 861466.

Heath, R. (2023). *China races ahead of the U.S. on AI regulation*. Anxious. https://www.axios.com/2023/05/08/china-ai-regulation-race

Henderson, P. (2020). *Law, policy and AI update*. Stanford University Human Centred AI. https://hai.stanford.edu/news/law-policy-ai-update-china-requires-ai-watermarks-chatgpt-wont-make-it-us-courtrooms

IndiaAI. (n.d) *AI and intellectual property rights*. India. https://indiaai.gov.in/ai-standards/ai-and-intellectual-property-rights

Jayal, A., McRobert, A., Oatley, G., & O'Donoghue, P. (2018). *Sports analytics: Analysis, visualisation and decision making in sports performance*. Routledge.

Jha, R. (2020). Sports law-problems and perspective. *Supremo Amicus, 18*, 717.

Johnson, M. (2023). *What are the AI's advantages and drawbacks of AI in the sports industry?* Sports Edtv. https://sportscdtv.com/blog/what-are-artificial-intel-s-sports-benefits-and-drawbacks

Kaur, J. (2023). *Comprehending ethical AI challenges*. Xenonstack. https://www.xenonstack.com/blog/ethical-ai-challenges-and-architecture#:~:text=Ethical%20AI%20should%20follow%20principles,%2C%20society%2C%20and%20the%20environment

Kemp, R. (2018). *Legal aspects of artificial intelligence (v2.0)*. Cyberspace Lawyer NL, 2.

Liao, S. M. (Ed.). (2020). *Ethics of artificial intelligence*. Oxford University Press.

Madiega, T. A. (2019). *EU guidelines on ethics in artificial intelligence: Context and implementation*. EPRS.

McCabe, A., & Trevathan, J. (2018). Artificial intelligence in sports prediction. In *Fifth international conference on information technology: New generations* (pp. 1194–1197). IEEE.

Mileva, G. (2023). *The rise of AI*. Influencer Marketing Hub. https://influencermarketinghub.com/ai-digital-transformation/#toc-2

Morgulec, E., Azar, O. H., & Lidor, R. (2018). Sports analytics and the big-data era. *International Journal of Data Science and Analytics, 5*, 213–222.

Naik, N., et al. (2022). Legal and ethical consideration in artificial intelligence in healthcare: Who takes responsibility? *Frontiers in Surgery, 9*, 862322. https://www.frontiersin.org/articles/10.3389/fsurg.2022.862322/full

Neznamov, A. V. (2019). Legal aspects of the implementation gf the national strategy for the development of artificial intelligence until 2030. *Courier of Kutafin Moscow State Law University (MSAL), 12*(64), 82–88.

Niziol, K. (2021). The challenges of consumer protection law connected with the development of artificial intelligence on the example of financial services (chosen legal aspects). *Procedia Computer Science, 192*, 4103–4111.

Novachkov, H., & Baca, A. (2013). Artificial intelligence in sports on the example of weight training. *Journal of Sports Science and Medicine, 12*(1), 27.

Olfers, M. (2009). *Sport and competition law: An interesting twosome*. On the Ball.

Olivia, D. (2020). Legal aspects of artificial intelligence on automated decision-making in Indonesia: Lessons from the European Union, the United States, and China. *Lentera Hukum, 7*, 301.

Parrish, R. (2003). Towards a theory of EU sports law and policy. In *Sports law and policy in the European Union* (pp. 23–60). Manchester University Press. http://www.jstor.org/stable/j.ctt155jc4s.6

Peterson, J. (2023). *AI in sports*. Witty Sparks. https://wittysparks.com/artificial-intelligence-in-sport-industry/

Petropolouos, G. (2018). *The dark side of AI*. The Bruegel. https://www.bruegel.org/blog-post/dark-side-artificial-intelligence-manipulation-human-behaviour

Ramkumar, P. N., Luu, B. C., Haeberle, H. S., Karnuta, J. M., Nwachukwu, B. U., & Williams, R. J. (2022). Sports medicine and artificial intelligence: A primer. *The American Journal of Sports Medicine, 50*(4), 1166–1174.

Ratiu, O. G., Badau, D., Carstea, C. G., Badau, A., & Paraschiv, F. (2010). Artificial intelligence (AI) in sports. In *Proceedings of the 9th WSEAS international conference on Artificial intelligence, knowledge engineering and data bases* (pp. 93–97). World Scientific and Engineering Academy and Society (WSEAS).

Rigano, C. (2018). *Using AI to understand criminal justice needs*. National Institute of Justice. https://nij.ojp.gov/topics/articles/using-artificial-intelligence-address-criminal-justice-needs

Rodrigues, R. (2020). Legal and human rights issues of AI: Gaps, challenges and vulnerabilities. *Journal of Responsible Technology, 4*, 100005. https://doi.org/10.1016/j.jrt.2020.100005

Rosen, H. (2023). Understanding how AI incorporates transformation. *Forbes*. https://www.forbes.com/sites/forbesbusinesscouncil/2023/03/16/understanding-how-to-effectively-incorporate-digital-transformation-and-ai/?sh=2984b775179e

Schreck, M., Gomez, M., & Charkoudian, S. G. (2023). *US AI regulations: Watchlist*. Goodwin. https://www.goodwinlaw.com/en/insights/publications/2023/04/04_12-us-artificial-intelligence-regulations

Smerdov, A., Somov, A., Burnaev, E., et al. (2022). AI-enabled prediction of video game player performance using the data from heterogeneous sensors. *Multimedia Tools and Applications, 82*, 11021–11046. https://doi.org/10.1007/s11042-022-13464-0

Vakkuri, V., & Abrahamsson, P. (2018). The key concepts of ethics of artificial intelligence. In *2018 IEEE international conference on engineering, technology and innovation (ICE/ITMC)* (pp. 1–6). IEEE.

Vakkuri, V., Kemell, K. K., Tolvanen, J., Jantunen, M., Halme, E., & Abrahamsson, P. (2022, June). How do software companies deal with artificial intelligence ethics? A gap analysis. In *Proceedings of the 26th International Conference on Evaluation and Assessment in Software Engineering* (pp. 100–109).

WeForum. (2023). *The European Union's AI Act, 2023*. https://www.weforum.org/agenda/2023/06/european-union-ai-act-explained/

Xian, L. (2010). Artificial intelligence and modern sports education technology. In *2010 International conference on artificial intelligence and education (ICAIE)* (pp. 772–776). IEEE.

Zhang, H. (2020). Value and development ideas of the application of artificial intelligence in sports training. *Journal of Physics: Conference Series, 1533*(3), 032049.

Enhancing Player Experience Through AI-Powered Wireless Sensor Networks: A KNN Algorithm Approach for Tracking Daily and Sports Activities

M. Narendran, R. Swarna Teja, K. Sumithra Devi, S. Gayathri, and S. Gayathri

1 Introduction

Ensuring an athlete is in the most significant case scenarios to exercise involves monitoring their training regimen. Coaches may personalize training plans and enhance the prescription of training load and recovery owing to monitoring. Supervision helps athletes become more conscious and conscientious, increasing their sense of ownership over their health.

Qian (2021) study's purpose is to investigate the use of wireless sensors in sporting events, train E-sports competitors, and raise competitors' technical and technological proficiency by utilizing wireless sensor networks. He et al. (2022) used wireless sensor networks with AI to diagnose metal oxide arrester (MOA) faults. The challenges of learning, monitoring, and certifying the next generation of intelligent wireless networks utilizing artificial intelligence/machine learning (AI/ML) are discussed by Raghothaman (2021). This work aims to alter the conventional teaching approach using human-computer interaction, which communicates utilizing language and actions, and gesture data. The objective of the project by Wang et al. (2019) is the evolution of fifth-generation cellular networks, and the authors of this study have also explored open problems and difficulties with spectrum sensing in 5G.

M. Narendran (✉) · K. Sumithra Devi
KCG College of Technology, Karapakkam, Chennai, Tamil Nadu, India

R. Swarna Teja
Dhanalakshmi Srinivasan College of Engineering and Technology, Chennai, Tamil Nadu, India

S. Gayathri
Mohamed Sathak A.J. College of Engineering, Chennai, Tamil Nadu, India

S. Gayathri
TJ Institute of Technology, Chennai, Tamil Nadu, India

A. Mansurali et al. (eds.), *Sports Analytics*,
https://doi.org/10.1007/978-3-031-63573-1_11

1.1 Background and Motivation

This research explores the historical context and underlying issues that emphasize the necessity of AI-powered wireless sensor networks in sports analytics. This study examines the progression of player monitoring methodologies and underscores the constraints associated with conventional methods. In addition, we explore the potential of integrating artificial intelligence (AI) with wireless sensor networks (WSNs) as a promising avenue to address these constraints and facilitate a paradigm shift toward data-centric optimization of sports performance.

1.2 Problem Statement

The domain of sports and athletics is seeing a significant shift in perspective as it increasingly embraces the utilization of data-driven insights to boost performance. Nevertheless, conventional player monitoring systems fail to deliver precise, up-to-date, and all-encompassing monitoring of daily and athletic endeavours. Current methodologies need help in accurately capturing the many complexities of players' motions, resulting in imprecise data and delayed provision of feedback. Furthermore, the challenge of recognizing dynamic activity patterns in real time presents an additional barrier to the ability to produce practical insights. These limitations impede coaches, trainers, and athletes from making informed judgments that depend on prompt and dependable data. Hence, there is a pressing need for an intelligent system that can effectively tackle these challenges by utilizing AI-powered wireless sensor networks.

1.3 Objective of the Study

- This chapter seeks to develop and implement an advanced player monitoring system that integrates artificial intelligence (AI) techniques and wireless sensor networks. The system's primary goal is to enable real-time tracking of individuals' daily and athletic activities.
- The K-nearest neighbours (KNN) algorithm can accurately differentiate and categorize various activity patterns, hence assuring precise activity tracking and analysis.
- The objective of this chapter is to improve the player experience by providing prompt and thorough information on performance metrics. This performance will enable athletes to make informed decisions regarding their training and overall performance optimization.
- The user's text consists of a bullet point symbol. This study aims to assess the existing system's ability to improve the efficiency of training programs by identifying areas that need improvement and offering personalized recommendations based on individual performance data.

1.4 Contribution of the Study

- The chapter proposes a new combination of artificial intelligence (AI) techniques with wireless sensor networks (WSNs) to overcome the limitations of traditional player monitoring systems. This integration provides enhanced precision and efficiency in monitoring tasks, offering a possible resolution to current challenges.
- The current chapter demonstrates an advanced strategy for detecting real-time activities, which surpasses conventional methods by utilizing the K-nearest neighbours (KNN) algorithm. This approach allows for prompt and precise surveillance of player actions.
- The designed system offers coaches, trainers, and players the chance to obtain extensive performance insights. This enables the utilization of data-driven methodologies in decision-making for customized training and performance optimization.
- The advancement of sports technology is driven by the integration of cutting-edge technologies, which aim to enhance the industry. This integration showcases the potential of artificial intelligence and wireless sensor networks for improving player experiences and performance results.

2 Literature Review

In the study by Gao (2022), the Decision Tree Algorithm (DTA) is employed, and the system enables the robot to communicate with the network to instruct athletes in badminton and educate the athletes on alternative techniques. The study by Wang and Zhu (2021) uses wireless sensor networks to track actual human activity during national sporting events, extract the players' personality patterns, and then broadcast those tendencies again to the sportsmen on time. The authors of this study investigated the advantages of convolutional neural networks over other machine learning techniques. To improve performance, Wang et al. (2020) recommended using a deep multi-scale convolutional neural network (DMCNN) for network intrusion detection. This study proposes two steps in the paradigm: intrusion detection and intrusion prevention. The proposed model categorizes the data after learning pertinent feature representations from many labelled data. This study uses convolutional neural networks to shield WSNs against intrusion. The wireless sensor network dataset (WSNDS) is used to assess and test the effectiveness of the suggested system as presented by Chandre et al. (2022). This study proposes a hybrid intrusion detection system (IDS) that combines the benefits of rule-based and machine learning-based approaches. More specifically, the authors used machine learning approaches to achieve a high detection rate while retaining a low computing demand by balancing the detection with a rule-based component by Zhang and Ma (2022). The authors Zhang and Ma (2022) employed a deep learning (DL)-based classification model for intrusion detection in MANET. Feature selection algorithms are used to clear out extraneous and repetitive data since the features in the data significantly impact classification accuracy as proposed by Murugan and Jeyakarthic (2019). This study proposes, constructs, and evaluates a deep convolutional neural network-based intrusion detection system (DCNN-IDS).

Altunay and Albayrak (2021) described the feature-based intrusion detection system developed using a convolutional neural network to thwart cyber-attacks. Khan (2021) created a convolutional recurrent neural network (CRNN)-based hybrid ID framework that anticipates and categorizes detrimental attacks in the network. Albattah and Rassam (2022) proposed a model that uses correlations between the numerous physiological data variables utilizing hybrid convolutional long short-term memory (ConvLSTM) techniques to detect fundamental point abnormalities and contextual anomalies in the enormous data stream of WBAN. This study proposed intrusion detection methodology is based on CNN-based feature selection, CNN-based feature reduction, and random forest-based classification. Compared to several recent investigations, the recommended technique exhibits its usefulness in high accuracy and high detection rate as projected by Bisen and Sharma (2020). To boost security and enhance the detection capabilities of hostile infiltration behaviour in a wireless network (ICNN), this research offers a wireless network intrusion detection technique based on enhanced convolutional neural networks as proposed by Yang and Wang (2019). Song et al. (2020) developed the DCNN model to provide superior detection performance while minimizing the unnecessary complexity in the Inception-ResNet model's design. It was adjusted for the CAN bus's data flow. Zhong et al. (2011) first use the random forest strategy to choose the essential qualities; eliminating irrelevant or unneeded inputs simplifies the task and enables quicker and more precise identification. Second, a conventional NN model trains and detects intrusions using the selected key characteristics. Karthiga and Santhi (2021) employed the artificial bee colony and the grey wolf optimizer to optimize clustering, enhance routing, and extend network lifespan. This study's approach for automatically detecting the distinct ECG segments uses a convolutional neural network.

2.1 Challenges and Limitations of the Activity Monitoring

Table 1 represents the existing challenges and limitations with the monitoring of the players' activities. The key methods include manual data collection, subjective analysis, lack of granularity to capture-specific movements, and more that incorporates privacy concerns in monitoring the sportsperson.

However, from Table 1 that specifies the limitations for monitoring the sportsperson activities, Table 2 is elaborated to specify the challenges and limitations that are faced in wireless sensor networking environment. From the traditional model of manual data collection for analysis, recent years there is a massive development in the wireless technologies leading to wearable devices for sequential monitoring of the sportsperson.

Table 1 Limitations and challenges of the sportsperson's activity monitoring

S. no	Challenge/ limitation	Description
1	Manual data collection	Numerous conventional techniques are dependent on human data collection, hence giving rise to the possibility of errors and inconsistencies
2	Subjective analysis	The subjectivity of human observers can influence the accuracy and consistency of data interpretation
3	Lack of granularity	Traditional approaches may not possess the level of detail required to record complex movements and subtle indicators of performance accurately
4	Limited real-time monitoring	The capacity to optimize performance is hindered by the delay or infrequency of data collecting, as it limits the timely provision of feedback
5	Complexity of analysis	Accurate analysis of complex movement patterns necessitates the utilization of advanced algorithms, a capability that may be lacking in older methods
6	Invasive equipment	Several techniques employ intrusive apparatus that could impact players' comfort and innate mobility
7	Privacy concerns	The privacy of players may be jeopardized when intrusive monitoring techniques are utilized
8	Data overload	Conventional approaches may yield a substantial volume of data, posing difficulties in extracting significant insights

Table 2 Challenges and limitations of sportsperson monitoring with wireless sensor networks

S. no	Challenge/ limitation	Description
1	Limited precision	Wireless sensor networks (WSNs) may encounter challenges regarding accuracy due to factors such as sensor noise, signal interference, and measurement mistakes
2	Data latency	Delays in the transmission and processing of data may impede the ability to monitor in real time and provide prompt feedback
3	Energy efficiency	Specific wireless sensor networks (WSNs) encounter difficulties sustaining energy-efficient operations, which impact sensor node durability
4	Scalability	The issue of network scalability in data management becomes a barrier as the quantity of sensor nodes grows
5	Coverage and placement	The complexity lies in ensuring complete coverage and proper placement of sensor nodes to acquire pertinent data
6	Data interference	Interference can pose a potential challenge to wireless communications, hence affecting the integrity and dependability of the acquired data
7	Signal transmission reliability	The occurrence of unreliable data transfer has the potential to result in the loss or corruption of data, hence impacting the overall accuracy of monitoring processes
8	Complex data processing	Efficient algorithms and resources are necessary for effectively handling and processing a substantial amount of data originating from various sensors

3 Artificial Intelligent Techniques in Sports Activity Recognition

The field of sports activity recognition has been significantly transformed by the advancements in artificial intelligence (AI), which have facilitated precise and automated analysis of players' motions. Machine learning methodologies have been recognized as highly effective instruments for discerning significant patterns from sensor data. These strategies leverage the capacity to acquire knowledge from data, adjust to different circumstances, and offer immediate insights for enhancing training and performance.

3.1 *Machine Learning and Pattern Recognition*

Machine learning is essential in sports activity recognition as it facilitates automating the intricate pattern-learning procedure from sensor data. Algorithms like decision trees, support vector machines, and neural networks are utilized to identify and distinguish unique movement patterns correlated with various activities. The algorithms undergo training using labelled datasets that include various activities, enabling the models to generalize and effectively categorize novel occurrences as highlighted by Chen and Guestrin (2016).

Machine learning also enables the extraction of complex properties from unprocessed sensor data, hence facilitating the identification of nuanced actions that may encompass several bodily movements. Machine learning approaches significantly reveal insights not readily evident through manual observation alone, achieved through feature extraction and pattern recognition as analysed by Zhang (2000).

3.2 *Role of K-Nearest Neighbours (KNN) Algorithm*

The K-nearest neighbours (KNN) algorithm is pivotal in pattern recognition and activity classification. The K-nearest neighbours (KNN) algorithm holds considerable importance in sports activity recognition, primarily because of its straightforwardness and efficacy in finding comparable cases by measuring their proximity in the feature space. The KNN algorithm is based on the notion that data points exhibiting similar features will likely be assigned to the same activity class as specified by Cover and Hart (1967).

The KNN algorithm entails the retention of labelled training data and the subsequent comparison of the attributes of an input data point with those of the training instances. The approach employs a method to determine the "K"-nearest neighbours to the given data point and subsequently provides a class label based on the predominant class within this set of neighbours. The KNN algorithm suits scenarios with non-linear decision boundaries and various activities as explained by Anguita et al. (2023).

4 Methodology

This research considers the evaluation of the existing convolutional neural network (CNN), deep convolutional neural network (D-CNN), and genetic algorithm (GA) with the proposed K-nearest neighbour (KNN) by implementing the model in the AI-WSN-based technology. The algorithm is implemented in the technology to collect data about the activities of the sportsperson with the aid of wearable sensors. The wearable sensors may include smart watches and other similar gadgets. Later during the indoor training sessions and the competition events, monitoring the sportsperson, starting from food intake, practicing outdoors, and others, are mandatory. The activity will be monitored continuously through the sensors that provide timely action by the coach in taking further decisions regarding the training and other supplementary activity for the players of any event. This model also supports making decisions on whether the players are undergoing the expected postures to achieve in the events participated. AI model supports the data collection with the wireless sensor devices, which are then connected to the wireless sensor networks. The KNN model is deployed to perform decision-making activities such as alerting the player about practice and informing the coach or the trainer if there is any difference in the regular bodily movement. Figure 1 represents the working phases of the proposed system.

The algorithmic format for the feature selection by the models is presented subsequently. Table 3 is the algorithmic notation for the CNN model in feature selection Table 3.

Table 4 represents the algorithmic representation of D-CNN model for feature selection in the activities of the sportsperson.

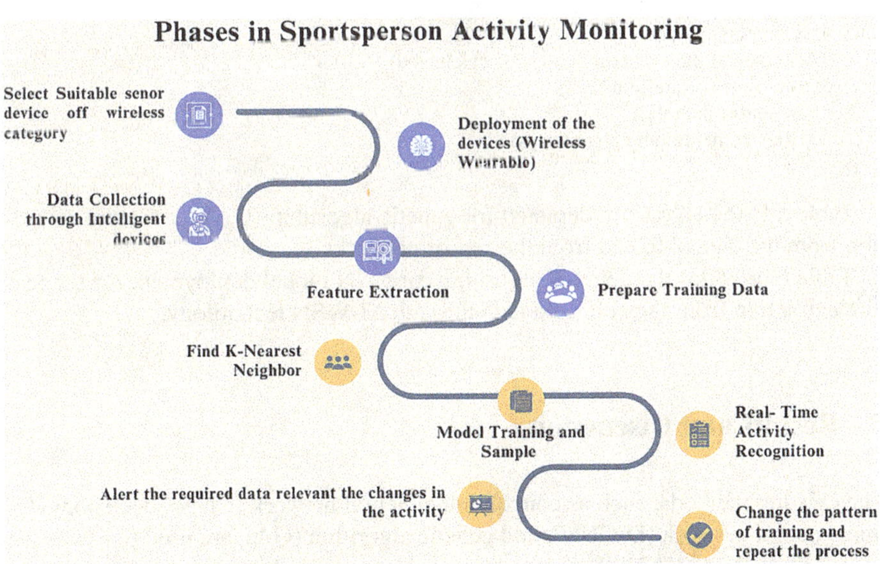

Fig. 1 Working phases of the proposed model

Table 3 CNN model algorithm

Algorithm for CNN model
Step 1: Collect and pre-process data … (Data collection and pre-processing) Step 2: Initialize CNN L: Number of layers W^l: Weights of the layer l b^l: Biases of the layer l a^{l-1}: Activation of previous layer F^l: Filter size of the layer l P^l: Padding of the layer l S^l: Stride of the layer l N: Number of training data points D^l: Depth of the layer l H^l and W^l: Height and width of the layer l X: Set of normalized feature vectors Y: Corresponding activity labels Step 3: Train model For each epoch: For each training example (x, y) Forward propagation: $z^l = W^l * a^l + b^l$ $a^l = \text{activation}\ (z^l)$ Compute loss: $L = \text{loss}(a^{(L)}, y)$ Back-propagation: $da^{(L)} = \text{lossDeriv}(a^{(L)}, y)$ $dz^{(L)} = da^{(L)} * \text{activateDeriv}(z^{(L)})$ … Update weights and biases: $W^{(l)} = W^{(l)} - \text{learnRate} * dW^l$ $b^{(l)} = b^{(l)} - \text{learnRate} * db^l$ Step 4: Evaluation model … (Accuracy of the evaluation model is calculated) Step 5: Real-time recognition For new sensor data x_{new}: … (Repeat the training phase steps)

Table 5 is the algorithm depicted for genetic algorithm (GA) for feature extraction from the data obtained from the sensor networks.

Table 6 provides the detailed algorithm for KNN model deployment for the feature extraction from the collected data through AI-WSN technology.

5 Results and Discussion

The existing methods, such as convolutional neural network (CNN), deep convolutional neural network (D-CNN), and genetic algorithm (GA), are used to assess the proposed methods KNN. The parameters include throughput, encryption, decryption, specificity, efficiency, and MSE.

Table 4 D-CNN model algorithm

Algorithm for D-CNN model
Step 1: Collect and pre-process data
... (Data collection and pre-processing)
Step 2: Initialize CNN
L: Number of layers
W^l: Weights of the layer l
b^l: Biases of the layer l
a^{l-1}: Activation of previous layer
F^l: Filter size of the layer l
P^l: Padding of the layer l
N: Number of training data points
D^l: Depth of the layer l
H^l and W^l: Height and width of the layer l
X: Set of normalized feature vectors
Y: Corresponding activity labels
Step 3: Train model
For each epoch:
For each training example (x, y)
Forward propagation:
$z^l = W^l * a^l + b^l$
$a^l = \text{activation}\ (z^l)$
Compute loss:
$L = \text{loss}(a^{(L)}, y)$
Back-propagation:
$da^{(L)} = \text{lossDeriv}(a^{(L)}, y)$
$dz^{(L)} = da^{(L)} * \text{activateDeriv}(z^{(L)})$
...
Update weights and biases:
$W^{(l)} = W^{(l)} - \text{learnRate} * dW^l$
$b^{(l)} = b^{(l)} - \text{learnRate} * db^l$
Step 4: Evaluation model
... (Accuracy of the evaluation model is calculated)
Step 5: Real-time recognition
For new sensor data x_{new}:
... (Repeat the training phase steps)

The most common unit of measurement for throughput is bits per second; however, data packets per second and data packets per time slot are other common alternatives. Figure 2 indicates the throughput for proposed and existing methods. Throughputs are higher in proposed methods and lower in existing methods.

Encryption time discusses and indicates the average time needed to encrypt input media content files. It is measured in seconds. When it comes to encryption, the amount of time it takes to encrypt a given media data is directly related to the input media content file size. Figure 3 indicates the encryption time. Encryption times are lower in proposed methods and higher in existing methods. Table 7 depicts the results of throughput and encryption time for existing and proposed methods.

Decryption refers to the process of converting encrypted data back to its original state. Reverse encryption is often used. It decodes encrypted data so that only authorized users are able to do so since decryption needs a secret key or password.

Table 5 GA model algorithm

Algorithm for GA model
Step 1: Collect and pre-process data ... (Data collection and pre-processing) Step 2: Initialize genetic algorithm P: Population size N: Number of features G: Number of generations considered M: Rate of mutation F: Fitness function Population: set of chromosomes initially considered as feature subsets Step 3: Evolutionary loop For each generation: For each chromosome (feature subset) in the population: Evaluate fitness: Fitness = F(Chromosome) Selection: Select parents based on fitness (e.g. roulette wheel or tournament selection) Crossover: Create offspring by combining genetic information from parents Mutation: Apply mutation to offspring's genetic information with a probability Replacement: Replace old population with new offspring population Step 4: Select best feature subset Select the chromosome with the highest fitness as the best feature subset Step 5: Real-time feature selection For new sensor data x_{new} received from AI-WSN: ... (Repeat the steps as evolutionary loop)

Table 6 GA model algorithm

Algorithm for KNN model
Step 1: Collect and pre-process data ... (Data collection and pre-processing) Step 2: Initialize KNN K: Number of nearest neighbors N: Number of training data points D: Dimensionality of feature space X: Set of normalized feature vectors Y: Corresponding activity labels Step 3: Classification loop For each test data point x_{new}: For each training data point x_i: Calculate Euclidean distance: $\text{EucDist}(x_{\text{new}}, x_i) = \sqrt{\sum x_{\text{new}}[j] - x_i[j]^2}, \text{for } j - 1 \text{ to } D$ Sort distances and select K-nearest neighbours: NearNeigh = K data point with smallest distance Predict activity based on majority vote among neighbours: PredictAct = model(NearNeigh) Step 4: Real-time recognition For new sensor data x_{new} received from AI-WSN: ... (Repeat the steps in classification loop)

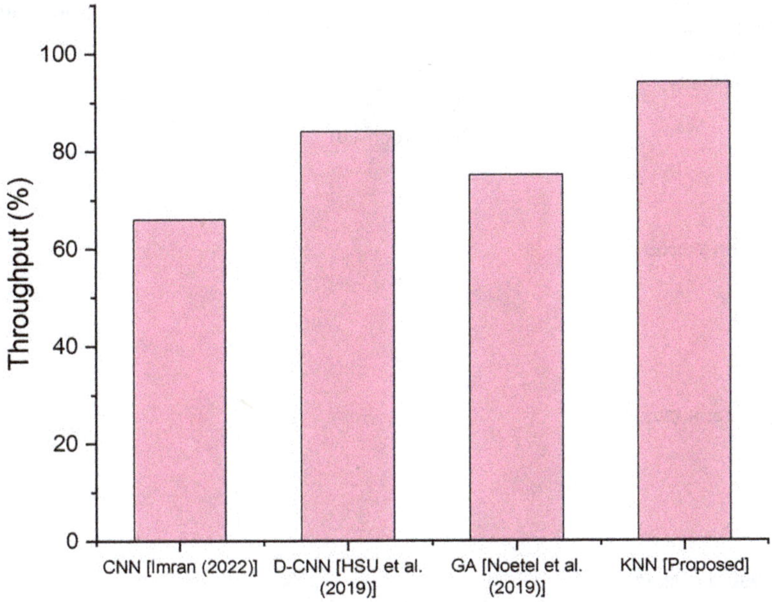

Fig. 2 Throughput

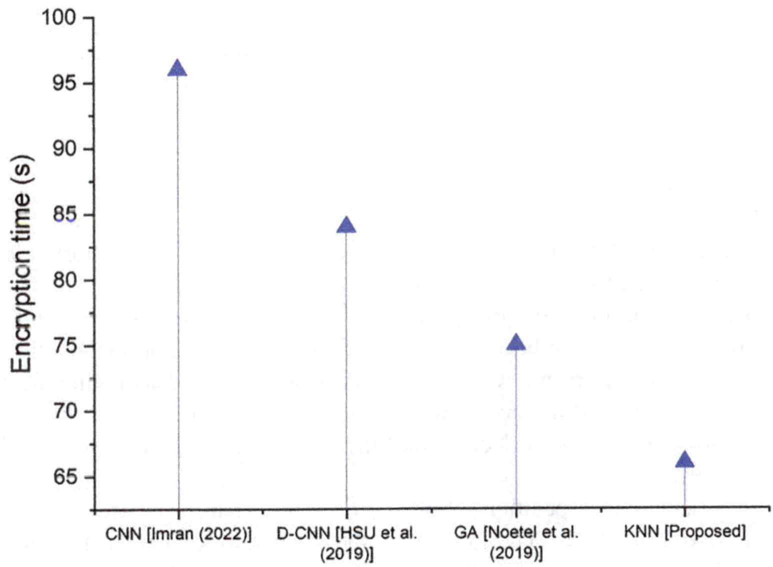

Fig. 3 Encryption time

Table 7 Results of throughput and encryption time for existing and proposed methods

	Throughput (%)	Encryption time (s)
CNN	66	96
D-CNN	84	84
GA	75	75
KNN (proposed)	94	66

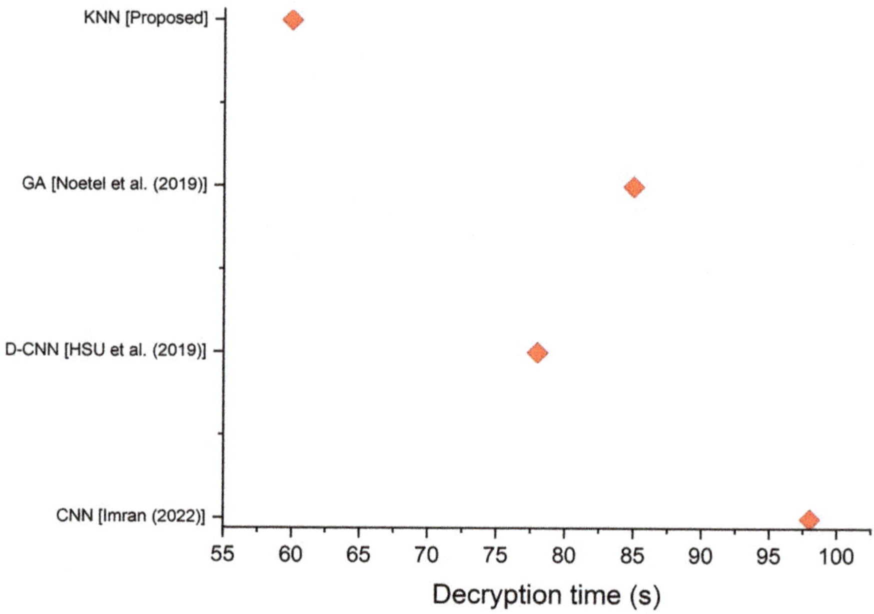

Fig. 4 Decryption time

Figure 4 depicts the decryption time. The proposed approaches have shorter decryption times compared to the existing methods' longer periods.

According to the principle of specificity, training should closely resemble the task at which one wishes to become more proficient. The more applicable the training is, the more of an impact it will have on the progress toward achieving the objectives. The fewer detailed the goals are, the longer it may take for to achieve them. Table 8 depicts the results of decryption time and specificity for existing and proposed methods.

Figure 5 depicts the specificity. Specificity is higher in proposed methods and lower in existing methods. The ratio of usable output to total input is a common metric for determining efficiency. This ratio may be stated mathematically using the formula $r = P/C$, where P is the quantity of useful output ("product") created in relation to C, which is the number of resources that were consumed.

The mean squared error is a statistical method that determines how closely a regression line corresponds to a given group of data points. It is a risk function with

Table 8 Results of decryption time and specificity for existing and proposed methods

	Decryption time (s)	Specificity (%)
CNN	98	78
D-CNN	78	84
GA	85	69
KNN (proposed)	60	95

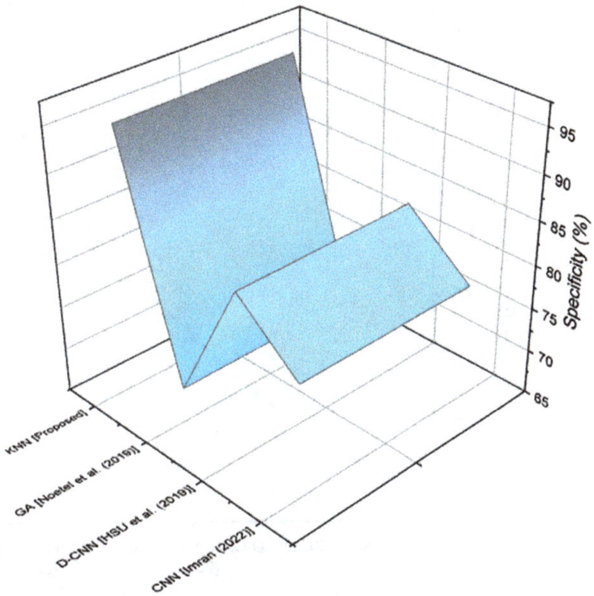

Fig. 5 Specificity

a value that corresponds to the anticipated squared error loss and its expected value. Table 9 depicts the results of efficiency and MSE for existing and proposed methods. From the below table, it can be observed that the proposed KNN model has outperformed other existing model in terms of efficiency and MSE by obtaining 93% and 61%, respectively. These results state the proposed model outperformed by marking a minimum increase of 8% with efficiency in comparison with GA and least error rate difference of 5% in comparison with the GA model.

Figure 6 indicates the efficiency. Efficiency is higher in proposed methods with 93% and lower in existing methods with 85%, 72%, and 62% for GA, D-CNN, and CNN, respectively.

Figure 7 depicts the MSE for proposed and existing works. The existing methods have a higher in MSE with 94% in CNN, 87% in D-CNN, and 76% with GA when compared to the proposed KNN with 61%.

Table 9 Results of efficiency and MSE for existing and proposed methods

	Efficiency (%)	MSE (%)
CNN	62	94
D-CNN	72	87
GA	85	76
KNN (proposed)	93	61

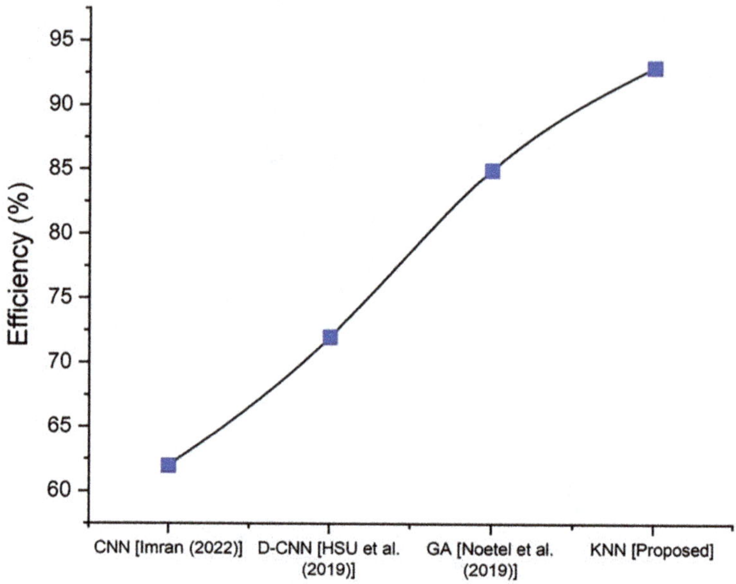

Fig. 6 Efficiency

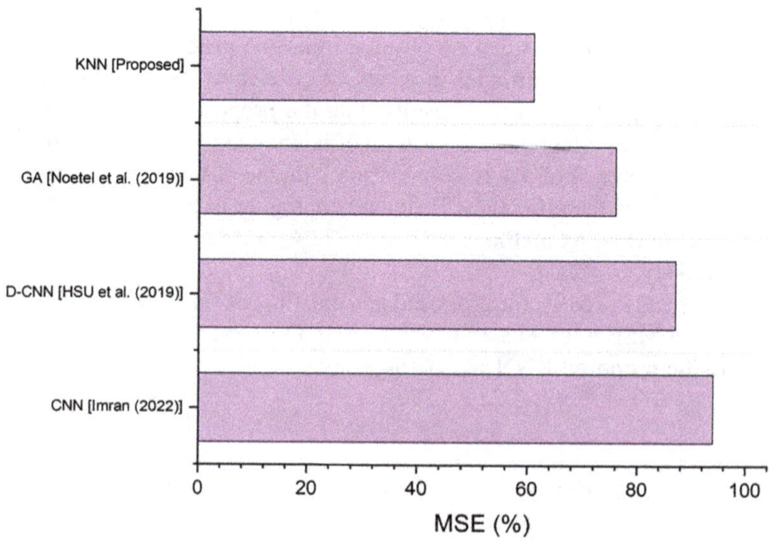

Fig. 7 MSE

6 Conclusion and Future Enhancement

6.1 Conclusion

The proposed research has illustrated the capacity to improve player engagement by incorporating AI-driven wireless sensor networks (WSNs) to monitor everyday and athletic pursuits. The efficacy of the K-nearest neighbours (KNN) algorithm in the classification and recognition of diverse activities has been well-documented, hence enhancing the comprehension of players' motions and actions in a holistic manner. The AI-WSN system that has been installed not only provides real-time monitoring capabilities but also enables the generation of data-driven insights regarding the performance and well-being of players.

Through the use of wireless sensor networks (WSNs) and artificial intelligence (AI) methodologies, many stakeholders within the realm of sports, such as players, coaches, and sports organizations, can derive many benefits. These advantages encompass precise monitoring and recording of physical activities, timely identification of irregularities, and the provision of tailored training regimens. The K-nearest neighbours (KNN) algorithm can classify actions by leveraging past data, presenting a non-intrusive and effective method for monitoring. Consequently, this strategy enhances player involvement and promotes injury prevention.

6.2 Future Enhancement

This research has effectively showcased the capabilities of AI-powered wireless sensor networks (WSNs) and the K-nearest neighbours (KNN) algorithm in sports activity tracking. However, some areas warrant more investigation and improvement in future research endeavours. Integrating various sensor modalities, such as accelerometers and heart rate monitors, can enhance the activity recognition and analysis dataset. This integration can yield more precise and resilient outcomes, improving the results' overall accuracy and reliability. The augmentation of the AI-WSN system to offer real-time feedback and alerts to players and coaches during training sessions has the potential to facilitate prompt modifications and enhance performance. The potential enhancement of temporal comprehension in activities can be explored by integrating deep learning approaches, specifically recurrent neural networks (RNNs) and long short-term memory (LSTM) networks.

The focus of this research is on the development of algorithms that possess the capability to adjust dynamically to novel or infrequent events. This adaptability enhances the flexibility and usefulness of the system in many sports and settings. The present research aims to analyse long-term performance to comprehensively examine the system's capacity and offer valuable insights into prolonged performance patterns and potential associations between various activities and player outcomes.

This chapter discusses the importance of privacy and data security in the context of player privacy and data security. It emphasizes the need to manage sensitive information responsibly and ethically.

References

Albattah, A., & Rassam, M. A. (2022). A correlation-based anomaly detection model for wireless body area networks using convolutional long short-term memory neural network. *Sensors, 5*, 1951. https://doi.org/10.3390/s22051951

Altunay, H. C., & Albayrak, Z. (2021). Network intrusion detection approach based on convolutional neural network. *European Journal of Science and Technology, 26*, 22–29. https://doi.org/10.31590/ejosat.954966

Anguita, D., Ghio, A., Oneto, L., Parra Perez, X., & Reyes Ortiz, J. L. (2023). A public domain dataset for human activity recognition using smartphones. In: *European symposium on artificial neural networks, computational intelligence and machine learning*. http://hdl.handle.net/2117/20897

Bisen, L., & Sharma, S. (2020). Convolutional neural network based intelligent network intrusion detection system. *Smart Moves Journal Ijoscience, 6*(9), 1–4. https://doi.org/10.24113/ijoscience.v6i9.317

Chandre, P. R., Mahalle, P., & Shinde, G. (2022). Intrusion prevention system using convolutional neural network for wireless sensor network. *IAES International Journal of Artificial Intelligence (IJ-AI), 11*(2), 504. https://doi.org/10.11591/ijai.v11.i2.pp504-515

Chen, T., & Guestrin, C. (2016). XGBoost. In *Proceedings of the 22nd ACM SIGKDD international conference on knowledge discovery and data mining*. ACM. https://doi.org/10.1145/2939672.2939785

Cover, T., & Hart, P. (1967). Nearest neighbor pattern classification. *IEEE Transactions on Information Theory, 1*, 21–27. https://doi.org/10.1109/tit.1967.1053964

Gao, S. (2022). The application of wireless network-based artificial intelligence robots in badminton teaching and training. *Computational Intelligence and Neuroscience, 2022*, 1–9. https://doi.org/10.1155/2022/3910307

He, T., Zhang, Z., Shen, P., Wei, M., & Zhang, Y. (2022). AI-based MOA fault diagnosis mechanism in wireless networks. *Wireless Networks, 30*, 1–12. https://doi.org/10.1007/s11276-022-03032-7

Karthiga, M., & Santhi, V. (2021). A wireless sensor network for remote detection of arrhythmias using convolutional neural network. *Wireless Networks, 4*, 1349–1360. https://doi.org/10.1007/s11276-021-02825-6

Khan, M. A. (2021). HCRNNIDS: Hybrid convolutional recurrent neural network-based network intrusion detection system. *Processes, 5*, 834. https://doi.org/10.3390/pr9050834

Murugan, S., & Jeyakarthic, M. (2019). Optimal deep neural network based classification model for intrusion detection in mobile ad hoc networks. *Journal of Advanced Research in Dynamical and Control Systems, 10*(Special Issue), 1374–1387. https://doi.org/10.5373/jardcs/v11sp10/20192983

Qian, F. (2021). Simulation training of E-sports players based on wireless sensor network. *Wireless Communications and Mobile Computing, 2021*, 1–10. https://doi.org/10.1155/2021/9636951

Raghothaman, B. (2021). Training, testing and validation challenges for next generation AI/ML-based intelligent wireless networks. *IEEE Wireless Communications, 6*, 5–6. https://doi.org/10.1109/mwc.2021.9690485

Song, H. M., Woo, J., & Kim, H. K. (2020). In-vehicle network intrusion detection using deep convolutional neural network. *Vehicular Communications, 21*, 100198. https://doi.org/10.1016/j.vehcom.2019.100198

Wang, Z., & Zhu, D. (2021). Sports monitoring method of national sports events based on wireless sensor network. *Wireless Communications and Mobile Computing, 2021*, 1–13. https://doi.org/10.1155/2021/5739049

Wang, D., Song, B., Chen, D., & Du, X. (2019). Intelligent cognitive radio in 5G: AI-based hierarchical cognitive cellular networks. *IEEE Wireless Communications, 3*, 54–61. https://doi.org/10.1109/mwc.2019.1800353

Wang, X., Yin, S., Li, H., Wang, J., & Teng, L. (2020). A network intrusion detection method based on deep multi-scale convolutional neural network. *International Journal of Wireless Information Networks, 4*, 503–517. https://doi.org/10.1007/s10776-020-00495-3

Yang, H., & Wang, F. (2019). Wireless network intrusion detection based on improved convolutional neural network. *IEEE Access, 7*, 64366–64374. https://doi.org/10.1109/access.2019.2917299

Zhang, Z. (2000). A flexible new technique for camera calibration. *IEEE Transactions on Pattern Analysis and Machine Intelligence, 11*, 1330–1334. https://doi.org/10.1109/34.888718

Zhang, L., & Ma, D. (2022). A hybrid approach toward efficient and accurate intrusion detection for in-vehicle networks. *IEEE Access, 10*, 10852–10866. https://doi.org/10.1109/access.2022.3145007

Zhong, S. H., Huang, H. J., & Chen, A. B. (2011). An effective intrusion detection model based on random forest and neural networks. *Advanced Materials Research, 267*, 308–313. https://doi.org/10.4028/www.scientific.net/amr.267.308

Predicting Indian Super League (ISL) Soccer Player Positions Using Personal and Performance Attributes

A. Mansurali, P. Raja, and V. Harish

1 Introduction

Sports have always been an integral part of Indian culture, and the rise of professional sports leagues has opened up new avenues for talented athletes to showcase their skills and abilities. The Indian Super League (ISL) is one of the most popular sports leagues in India, attracting a significant fan following and generating considerable revenue for the sports industry. The Indian Super League (ISL) is India's top professional soccer league, reflecting the country's growing interest in the sport. The ISL, founded in 2013, has quickly become one of the top-tier football leagues in the country, attracting fans and improving the sport. The league mixes domestic and international talent and offers a variety of footballing styles and methods. The ISL has effectively introduced a franchise-based format to Indian football, similar to other leagues. Notably, the league pits city franchises against each other for the title. The league's annual calendar includes group-stage matches, elimination rounds, and a grand finale to determine the season's winner. The ISL has attracted attention for its focus on developing local potential while drawing top foreign players and

A. Mansurali (✉)
Department of Management, Central University of Tamil Nadu,
Thiruvarur, Tamil Nadu, India
e-mail: mansurali@cutn.ac.in

P. Raja
Department of Tourism and Hospitality Management, Central University of Tamil Nadu,
Thiruvarur, Tamil Nadu, India
e-mail: rajaperumal@cutn.ac.in

V. Harish
PSG Institute of Management, PSG College of Technology, Coimbatore, Tamil Nadu, India
e-mail: harish@psgim.ac.in

A. Mansurali et al. (eds.), *Sports Analytics*,
https://doi.org/10.1007/978-3-031-63573-1_12

coaches to improve Indian football and promote competitiveness. This unusual blend electrifies supporters, making the ISL a key sports league in the country.

The Indian Super League (ISL) has emerged as a premier football competition in India, captivating the nation with its thrilling matches and showcasing a high level of talent on the field. As the league continues to evolve, the demand for advanced analytics and data-driven insights has become increasingly prominent. One crucial aspect of football analytics is the ability to predict player positions based on a combination of personal attributes and performance metrics. This chapter delves into the intriguing realm of predicting ISL player positions, exploring the intricate relationship between an individual player's characteristics and their on-field roles. However, with the increasing popularity of these leagues, the competition among teams to recruit the best players has become fierce, and identifying talented players has become a crucial aspect for sports teams and league managers. The chapter has an objective of building models to predict the position of soccer players based on their personal and performance attributes. Through this chapter, the goal is to provide practical implications for sports leagues and teams in India, helping them make data-driven decisions in talent identification and salary projections. The chapter's originality lies in its use of machine-learning algorithms and the integration of data. Overall, this chapter aims to improve the accuracy and effectiveness of talent identification methods in the Indian sports industry. The contributions of this chapter lie in its potential to enhance our understanding of the factors influencing player positions in the ISL and the broader context of football analytics. The predictive models developed herein could serve as valuable tools for coaches, analysts, and team management, aiding in strategic decision-making and talent acquisition. Moreover, the insights gleaned from this study may pave the way for future research in player development, tactical innovation, and the evolution of football analytics in the Indian context. As we embark on this exploration of data-driven insights in the context of Indian Super League soccer, the journey promises not only to unravel the intricacies of player positions but also to contribute meaningfully to the burgeoning field of football analytics. Through a meticulous blend of statistical rigor and machine-learning prowess, we aim to uncover the secrets hidden within the data, ultimately arming the football community with tools to decipher the beautiful game in new and unprecedented ways.

2 Background and Significance

The landscape of football has undergone a paradigm shift with the advent of advanced analytics and technology. In the ISL, where the competition is fierce and every tactical decision matters, understanding player positions is paramount. Traditionally, coaches relied on their intuition and experience to assign positions to players. However, with the wealth of data available today, there is an opportunity to enhance decision-making by leveraging statistical models to predict player positions accurately. The significance of predicting player positions lies in its potential to revolutionize team strategies, recruitment processes, and overall game dynamics. By deciphering the correlation between personal attributes, such as age, height, and playing style, and performance metrics, such as goals scored, assists, and defensive actions, we can gain valuable insights into the preferred positions for each player. This knowledge can empower coaches to optimize team formations, make informed substitutions, and devise strategies tailored to exploit opponents' weaknesses. This chapter aims to provide a comprehensive analysis of the factors influencing player positions in the ISL, drawing on a diverse set of personal and performance attributes. The scope of the study encompasses players from multiple seasons, ensuring a robust dataset that captures the evolution of the league and player dynamics over time. By identifying patterns and trends within the data, the goal is to develop predictive models capable of forecasting player positions with a high degree of accuracy. The chapter aims to develop effective predictive models for determining the position of soccer players by utilizing machine-learning algorithms such as multinomial logistic regression, support vector machines, random forest, and ensemble approaches. These models will be based on the players' personal and performance attributes. By providing valuable insights, this can assist sports leagues and teams in India in making well-informed choices about the recruitment and retention of skilled athletes. Ultimately, this can significantly contribute to the success and expansion of the sports business in India.

3 Literature Review

To contextualize the current study, a thorough examination of existing literature on football analytics and player position prediction is essential. Previous research in this field has explored various methodologies, ranging from traditional statistical analyses to sophisticated machine-learning algorithms. Understanding the strengths and limitations of these approaches will inform the design and implementation of our predictive models. Noteworthy studies have investigated the impact of physical attributes on player positions, highlighting the importance of factors such as height, speed, and body mass index. Additionally, performance metrics such as passing accuracy, defensive contributions, and attacking prowess have been scrutinized to unveil patterns in player roles. By building upon the findings of these studies, our chapter aims to contribute to the growing body of knowledge in football analytics, specifically within the context of the ISL.

The vast corpus of research on football analytics and player position prediction encompasses a broad range of methodologies, encompassing both conventional and contemporary approaches. The initial research conducted by Reep and Benjamin (1968) established the foundation for spatial analysis in football, with a specific emphasis on the complex positioning of players on the pitch. Gonçalves et al. (2012) provide an overview of some unanswered questions related to talent identification and specialization in sports. The authors highlight the need to determine the most appropriate age for specialization, as well as the importance of considering the athletes' physical, psychological, and sociological characteristics during the talent identification process. Buekers et al. (2015) reflect on the search for future champions and the difficulties associated with talent identification in sports. The authors emphasize the importance of considering a wide range of factors, such as physical and psychological attributes, family support, and environmental factors, when identifying young talents. They suggest that a comprehensive approach to talent identification that takes into account multiple factors can enhance the accuracy of talent identification and increase the likelihood of identifying future champions. Barraclough et al. (2022) provide a descriptive review of methodological approaches to talent identification in team sports. The authors suggest that a multidisciplinary approach that considers a wide range of factors is necessary for accurate talent identification. Further studies conducted by Bradley et al. (2019) and Bloomfield et al. (2019) highlighted the significant impact of physical characteristics, such as height and weight, on the allocation of players to specific positions. This research emphasized the idea that certain positions may be more suitable for players with particular physical traits, highlighting the growing understanding of how biology and on-field duties interact.

Machine-learning approaches have revolutionized football analytics. Link and Lang (2017) were the first to use machine-learning techniques to demonstrate the potential of forecasting player positions. Jamil et al. (2021) employed multiple machine-learning algorithms to classify professional men's football goalkeepers into elite and sub-elite categories. The authors suggest that machine-learning algorithms may provide an objective and efficient approach to talent identification. They highlight the importance of developing and validating reliable and valid assessment tools and the need to ensure that machine-learning algorithms are transparent, interpretable, and fair. Liu et al. (2020) enhanced this strategy by utilizing decision trees and ensemble methods to predict player roles by considering a combination of personal qualities and in-game activities. The incorporation of machine learning into football analytics not only improves the accuracy of predictions but also allows for the detection of intricate, nonlinear connections that standard statistical methods may fail to uncover. Simultaneously, scholars such as Lago-Peñas et al. (2016) have explored the complex correlation between performance indicators and player positions. By examining metrics such as passing accuracy, tackles, and defensive contributions, these studies offer a detailed comprehension of the complex and diverse aspects of player responsibilities. The focus on performance measures contradicts the traditional belief that player places are primarily based on physical characteristics, therefore emphasizing the flexible and responsive nature of football methods.

Garris and Wilkes (2017) focus on soccer athletes' salaries for the World Cup. They analyzed data from the 2014 World Cup and used econometric techniques to estimate the influence of various factors on the salaries of soccer athletes. The review highlights the importance of a player's age, experience, and position in determining their salary. The results suggest that strikers earn higher salaries than defenders and midfielders. Leite et al. (2021) highlight the importance of talent identification and development in sports performance. The authors provide an outline of the current state of research on talent identification and development and suggest several areas for future research. They emphasize the need to better understand the factors that contribute to success in different sports and at different levels of competition, as well as the importance of using evidence-based approaches to talent identification and development. An examination of the development of playing styles and their impact on positional dynamics has also been a central focus of research. Bradley et al. (2019) investigated the range of playing styles, illustrating how personal preferences in offensive and defensive tactics influence the fit for specific positions. Reep et al. (2008) highlighted the significance of playing styles in influencing team formations and tactical strategies. This study emphasizes the importance of going beyond a fixed comprehension of player positions and acknowledging the dynamic nature of roles within the always-evolving football environment.

Although these works significantly enhance the overall comprehension of football analytics, they have not extensively studied the unique context of the Indian Super League (ISL). The existing body of research on Indian football dynamics, as highlighted by Chatterjee (2018), reveals a notable deficiency, therefore presenting the Indian Super League (ISL) as a captivating and relatively unexplored area for further study. The distinctive attributes of the ISL, including its varied talent pool, cultural influences, and changing competitive environment, justify a thorough examination of player positions within this framework. The inclusion of more recent studies enhances the existing body of literature on football analytics. Pappalardo et al. (2019) examined the spatial characteristics of football, highlighting the significance of player mobility and positioning. The study offers valuable insights into how spatial factors influence player positioning, so introducing an additional level of intricacy to the current comprehension of on-field dynamics. This spatial analysis adds to the increasing acknowledgment that player positions are influenced not only by individual acts but also by the wider spatial context in which these activities take place. It is essential to incorporate a wide range of perspectives from many research to accurately anticipate the positions of players in the ISL, given the current football environment. The growing dependence on data-driven methodologies, combined with breakthroughs in machine learning, offers a chance to connect the divide between conventional and technology advancements in football analysis. The ISL, being a burgeoning football league, provides an optimal environment for conducting such an inquiry, owing to its distinctive attributes and the dynamic state of football in the Indian subcontinent.

The literature study on football analytics and player position prediction demonstrates the progression of methodology, starting from conventional spatial analysis and advancing to more complex machine-learning techniques. Player roles in football are shaped by physical features, performance metrics, playing styles, and spatial dynamics. Integrating these lessons into the specific setting of the Indian Super League offers a unique and demanding opportunity for research. By combining various studies, the ISL aims to gain insights into predicting player positions, which contributes to the ongoing discussion about the relationship between tradition and innovation in football analysis. These studies include the works of Bradley et al. (2019), Link and Lang (2017), Pappalardo et al. (2019), Liu et al. (2020), Chatterjee (2018), Reep et al. (2008), Bloomfield et al. (2019), and Lago-Peñas et al. (2016).

4 Methodology

The objective of this chapter is to utilize machine-learning algorithms to examine the performance and characteristics of players in ISL sports leagues. The goal is to construct prediction models that assess the influence of different variables, including age, region, performance data, and player attributes, on the player's position.

The methodology employed in this chapter involves a multifaceted approach to data collection, preprocessing, and modeling. The dataset will be meticulously curated to include relevant personal attributes and performance metrics for ISL players. Exploratory data analysis (EDA) will be conducted to uncover patterns and correlations within the data. Subsequently, predictive models will be developed using machine-learning algorithms, with a focus on accuracy, interpretability, and generalizability. The choice of machine-learning algorithms had been driven by the nature of the data and the complexity of the task. Classification models, such as multinomial logistic regression, support vector machines, random forest, and ensemble methods, have been explored to predict player positions.

The chapter methodology encompasses various steps, such as data gathering, data cleansing, feature engineering, model selection, model training, and model evaluation. Data cleaning includes the elimination of duplicate and missing data, as well as the handling of outliers and anomalies. Feature engineering entails the careful selection and manipulation of the pertinent features that are expected to have the greatest impact on the outcome variables.

The chosen machine-learning models were subsequently trained on the processed and converted data to forecast the desired outcome variables. The model evaluation procedure entails evaluating the accuracy, precision, and recall of the models by employing diverse performance metrics, including mean squared error, root mean squared error, and accuracy scores. Additionally, it seeks to develop predictive models that can inform decision-making in player selection and team composition. The outcomes of this chapter can be employed by team management, coaches, and players to improve their performance and optimize their potential.

5 Models Used

5.1 Random Forest

Random forest is an ensemble machine-learning algorithm that combines multiple decision trees to improve the accuracy of the model. It can be used for both classification and regression problems. The algorithm builds multiple decision trees on different subsets of the training data, and each tree is constructed using a random subset of the features. The final prediction is made based on the average or majority vote of the predictions made by the individual trees.

5.2 *Logistic Regression*

It is a machine-learning algorithm used for classification problems where the goal is to predict the probability of a binary outcome variable based on one or more input variables. It is a type of generalized linear model that uses a logistic function to model the relationship between the input and output variables. For regression problems, logistic regression is not applicable as the output variable is continuous, and the model would not be able to predict a range of values.

5.3 *Gradient Boosting*

Gradient boosting is a machine-learning algorithm that uses boosting to create an ensemble of decision trees. It combines multiple decision trees in a serial manner where each decision tree tries to correct the errors of the previous one. It works by assigning weights to the data points and then fitting a decision tree to the weighted data. The algorithm then uses gradient descent to minimize the errors and improve the accuracy of the model. Gradient boosting is typically used in scenarios where there is a large dataset and the goal is to achieve high accuracy in classification problems.

6 Data

This chapter relies on a large collection of player-centric data to analyze the Indian Super League (ISL). Multiple seasons are covered in this dataset, revealing league dynamics. Our analysis covers a large number of seasons from the ISL's 2014 launch to offer a representative and longitudinal knowledge of player performance trends. The dataset carefully selects domestic and international players who have contributed to the league's compelling story. This large dataset examines player traits and performance statistics to provide a detailed view of ISL player placements and outcomes. To contextualize our analysis, the dataset includes information on a large number of players, showing the diverse talents of the ISL across the covered seasons. Our machine-learning algorithms capture player behaviors across positions, age groups, locations, and salary bands due to their inclusivity.

The dataset for analysis includes both categorical and numerical variables that capture various dimensions of player performance in the Indian Super League (ISL). Categorical variables, such as player name, squad, position, and country, provide unique identifiers and contextual information. Numerical variables encompass player attributes and performance statistics, covering metrics like age, matches played, goals, assists, passing accuracy, tackles, salary, and overall score. This extensive dataset spans multiple seasons since the inception of the ISL in 2014, offering a comprehensive view of player dynamics. With data on diverse player attributes and performance metrics, including offensive and defensive statistics, the analysis aims to leverage machine-learning algorithms for predictive insights into player positions, yellow card likelihood, and overall performance in the ISL. This dataset, inclusive of a significant number of players and spanning various performance indicators, serves as a robust foundation for exploring the intricacies of the Indian sports landscape (Table 1).

Table 1 ISL data

Variables	Numerical vs. categorical	Explanation
Player name	Categorical	Name of the player
Squad	Categorical	Team name
Position	Categorical	The role that a player occupies on the field during a match
Age	Numerical	Age
Country	Categorical	Country of origin
Matches	Numerical	No matches played
Minutes	Numerical	The amount of time a player spends on the field during a match
Goals	Numerical	No. of times when the ball is successfully kicked or headed into the opposing team's net
Assists	Numerical	An assist is recorded when a player creates a decisive pass or play that directly directs to a goal being scored by their team
Goal conversion rate (%)	Numerical	The total number of shots they have taken is divided by the number of goals they have scored
Goals/game	Numerical	The average number of goals scored by a player or a team in a single match
Shots	Numerical	To measure the number of times a player or team attempts to score a goal by kicking the ball toward the opponent's goalpost
Crosses	Numerical	A pass that is targeted from the wide regions of the field to the center region of the opponent's field, to approach their goal, over a medium-to-long range distance
Goals conceded	Numerical	The number of goals that a team or goalkeeper has allowed the opposing team to score in a given period
Saves	Numerical	The number of times a goalkeeper successfully prevents a shot from entering the goal
Saves percentage	Numerical	Dividing the number of saves made by the goalkeeper by the number of shots faced by the opposing team
Goals conceded/game	Numerical	Dividing the total number of goals conceded by the team over the course of a season by the number of games played
Catches	Numerical	The number of times a goalkeeper successfully catches the ball during a match

(continued)

Table 1 (continued)

Variables	Numerical vs. categorical	Explanation
Passes	Numerical	The number of successful passes made by a player during a game or a season
Touches	Numerical	The number of times a player touches the ball during a game
Avg passes/game	Numerical	Average number of passes made by a player per game
Passing accuracy	Numerical	Percentage of successful passes made by a player in a game
Shots faced	Numerical	The number of shots attempted by the opposing team on a goal
Shots on target faced	Numerical	Measures the number of shots taken by the opposing team that are on target and that the goalkeeper has to save
Tackles	Numerical	Measures the number of times a player effectively steals the ball from the opposition
Interceptions	Numerical	The number of times a player intercepts or takes control of the ball from an opposing player
Clearances	Numerical	Measures the number of times a player or a team clears the ball out of their own half or penalty area to prevent the opposition from scoring
Free kicks	Numerical	No free kicks faced
Blocks	Numerical	Number of times a player or a team blocks a shot taken by the opposition team
Fouls	Numerical	Number of rule violations committed by a player during a match
Yellow cards	Numerical	Number of times a player has been cautioned by the referee
Red cards	Numerical	Number of times a player has been sent off the field of play and is no longer allowed to participate in the remainder of the match
Salary (pounds)	Numerical	Salary of players
Overall score	Numerical	A rating given to individual players based on their performance in a match or over the course of a season
Category (based on overall score)	Categorical	Level of performance or skill of a player based on their overall score

Source: www.Indiansuperleague.com

7 Preprocessing Techniques

7.1 Data Cleaning

One of the crucial stages in the data analysis process is data cleaning, which entails identifying and rectifying errors and discrepancies in the data. The quality of the data is essential for obtaining accurate and reliable results from data analysis. Python libraries such as Pandas provide useful functions for performing data-cleaning tasks. These functions include removing duplicate rows, handling missing values, and correcting data types. For example, the drop_duplicates() function had been used to remove duplicate rows in a DataFrame, while the fillna() function can be utilized to fill in missing values using specified methods such as mean, median, and mode. In addition to removing duplicates and filling in missing values, data cleaning also involves identifying outliers and deciding how to handle them using box plots, or statistical methods such as Z-score or IQR (interquartile range) methods. Depending on the background of the analysis, outliers can either be removed from the data or treated as missing values. Overall, data cleaning plays a vital role in the data analysis process, as it ensures that the data is reliable and accurate for further analysis.

7.2 Data Transformation

Data transformation helps to convert raw data into a more suitable format for further analysis. This step may involve a series of tasks such as cleaning the data, encoding categorical variables, scaling or normalizing numerical variables, and dealing with missing values. Data transformation aims to ensure that the data is appropriately formatted for analysis, thereby providing accurate insights. One of the primary tasks involved in data transformation is cleaning the data. This involves identifying and correcting errors or inconsistencies in the data. Python libraries such as Pandas offer a range of functions to perform data-cleaning tasks. Another important task in data transformation is encoding categorical variables. Categorical variables are variables that can take on a limited number of values or categories, such as gender and color. Encoding categorical variables is essential to convert them into numerical variables that can be used for analysis. Python libraries such as Pandas provide functions for encoding categorical variables, such as the get_dummies() function, which creates dummy variables representing the categories in the original variable. Scaling or normalizing numerical variables is another task involved in data transformation. This is important because variables may have different scales or units of measurement, which can affect the accuracy of the analysis. Python libraries such as NumPy provide functions for scaling and normalizing numerical data, such as the StandardScaler() function, which scales data to have zero mean and unit variance.

7.3 Feature Engineering

Feature engineering involves creating new variables or features from existing ones to enhance the performance and accuracy of the analysis. Python libraries, like Pandas, provide a variety of functions for creating new features that extract meaningful information from the data. For example, by combining multiple features, we can calculate the ratio of two numerical variables or extract information such as the month or day from a date feature. These new features can provide valuable insights into the data and help to build more accurate models. Another technique used in feature engineering is creating dummy variables, which is particularly useful when dealing with categorical variables. For example, in the case of the ISL, dummy variables for each team were created, indicating whether they won, lost, or drew a match. In addition to creating new features, feature engineering can also involve removing redundant or irrelevant features that do not contribute to the analysis. This can help to streamline the data and reduce the risk of overfitting, where the model is too closely fitted to the training data, resulting in poor performance on new data.

7.4 Data Standardization

Data standardization, also known as normalization, is a process of transforming numerical data so that it has a mean of 0 and a standard deviation of 1. This process involves subtracting the mean of the data from each data point and dividing by the standard deviation. In Python, the Scikit-learn library provides a function called StandardScaler that can be used for standardizing data. This function takes in a matrix or data frame and returns a scaled version of the input, where each column of the input data is standardized separately. Standardizing data is a crucial step in machine learning as it helps to make sure that all variables are on the same scale, preventing one variable from dominating others. Scikit-learn's StandardScaler function is particularly useful when working with datasets that contain variables with different units or scales. Standardizing the data ensures that each variable contributes equally to the analysis, leading to more accurate and reliable results. Machine-learning algorithms such as k-nearest neighbors and support vector machines benefit greatly from standardized data. Standardization helps to improve the accuracy of these algorithms as it puts all variables on the same scale, making them easier to compare and analyze. This makes it easier for the algorithm to identify important patterns and relationships in the data.

Objective

- To build a model to predict the position of a soccer player based on personal and performance attributes

Models Used Multinomial logistic regression, random forest, support vector machine, and gradient boosting classifier.

Dependent Variable *Position.*

The objective is to develop a model that predicts the position of a soccer player based on personal and performance attributes. The dependent variable for the model is POSITION, and the independent variables include various factors such as age, matches, minutes, goals, assists, saves, shots, tackles, fouls, yellow cards, red cards, and salary, among others. Four different machine-learning models, namely, multinomial logistic regression, random forest, support vector machine, and gradient boosting classifier, were used to develop the model. The independent variables also include the overall score and the country of the player, represented as binary variables. The objective is to achieve a high degree of accuracy and precision in predicting the position of a soccer player, which would be useful in the talent identification process.

8 Exploratory Data Analysis

EDA, or exploratory data analysis, is a vital step in the data analysis process that involves examining and understanding the patterns, trends, and distributions present in the data. EDA involved various techniques, including data visualization, summary statistics, and correlation analysis, to explore the data and uncover relationships between variables. These insights were then used to guide the development of machine-learning models and inform talent identification for ISL (Fig. 1).

The bar graph illustrates the influence of players' nation of origin on their overall score in the Indian Super League (ISL). Based on the graph, it is evident that players from Morocco and Fiji exhibit the greatest mean overall scores, whereas players from India demonstrate the lowest mean overall scores. Furthermore, the error bars represent the degree of variation in the data for each country, with Nigeria and France having the largest error bars.

From a business standpoint, this information might be valuable for managing teams and identifying talent. Teams can utilize this data to specifically focus on recruiting players from nations with superior average overall scores, such as

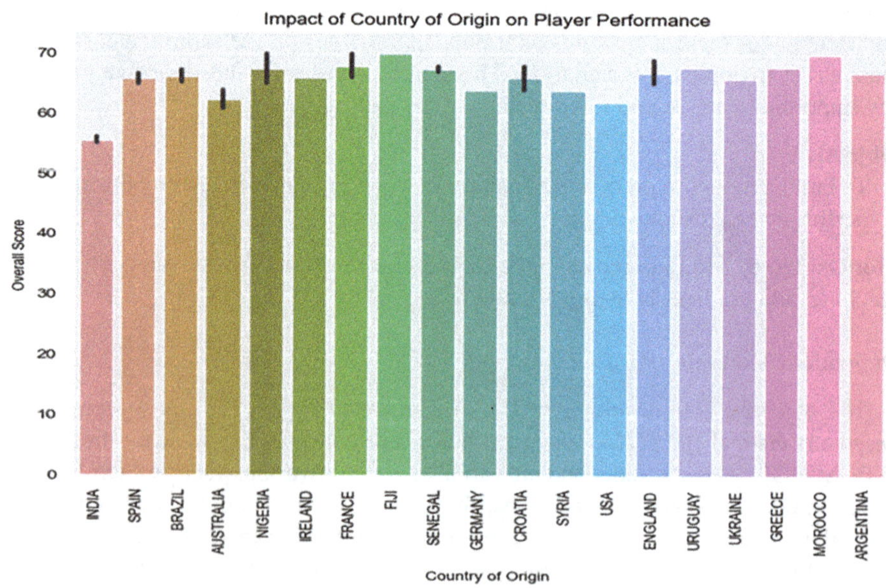

Fig. 1 Country and performance

Morocco and Fiji. Likewise, teams may opt to allocate less attention to players hailing from nations with lower mean scores, like India, unless there are extraordinary individual players. Moreover, the data regarding the fluctuation in scores among various countries might assist teams in making well-informed judgments while evaluating players from those nations, considering the possible hazards linked to their performance.

9 Impact of Age on Player Performance

The scatter plot depicting the relationship between the total score and age suggests the presence of a moderate positive connection between these two factors (Fig. 2). There is a positive correlation between age and overall score, meaning that as age increases, the overall score also tends to increase. The clustering of data points within the age range of 22.5–30 and the overall score range of 50–58 suggests that players within this age group are likely to excel in terms of their overall score. The line graph displays a linear trend line that depicts the correlation between age and the overall score. Within the framework of the scatterplot depicting the relationship between age and overall score, the trend line serves to elucidate the general pattern

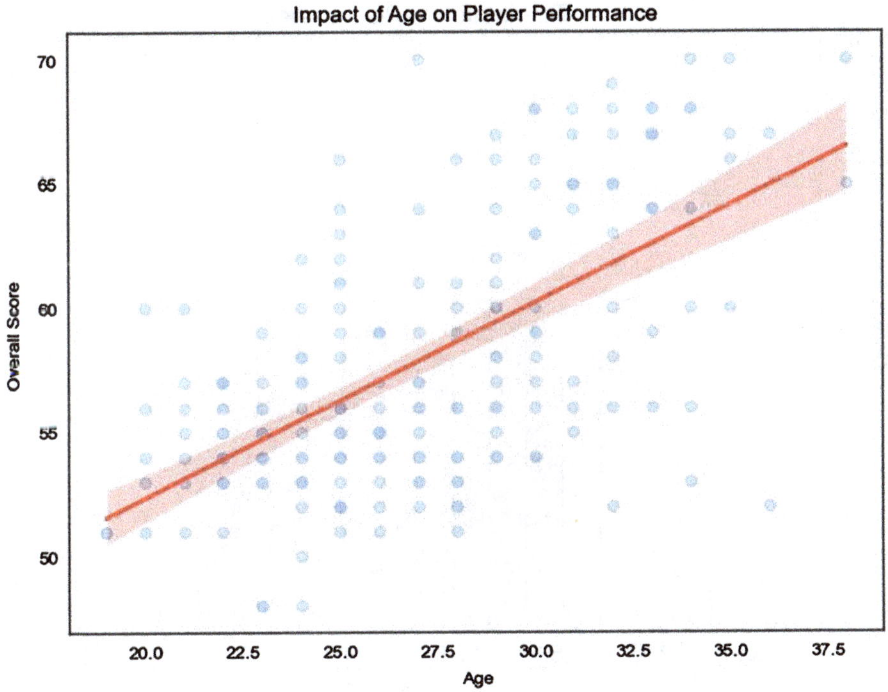

Fig. 2 Relationship between age and score

exhibited by the data. A positive slope signifies a direct correlation between age and overall score, implying that as the age of players increases, their total score also tends to increase.

From a commercial standpoint, this information might be utilized to shape recruitment and training efforts. Teams may opt to prioritize the recruitment and training of players aged between 22.5 and 30 to maximize their potential score. Moreover, this information can be utilized to develop customized training programs that concentrate on enhancing performance in particular domains that have a strong correlation with the overall score, such as physical fitness and technical abilities. This has the potential to finally result in enhanced team performance and triumph in the league.

10 Distribution of Player Age

A normal distribution of player ages indicates that the bulk of players are concentrated inside a specific age range, while there are fewer players at the extremes of the range (Fig. 3). Regarding this scenario, the distribution exhibits a prominent peak at approximately 25 years of age, indicating a higher concentration of players

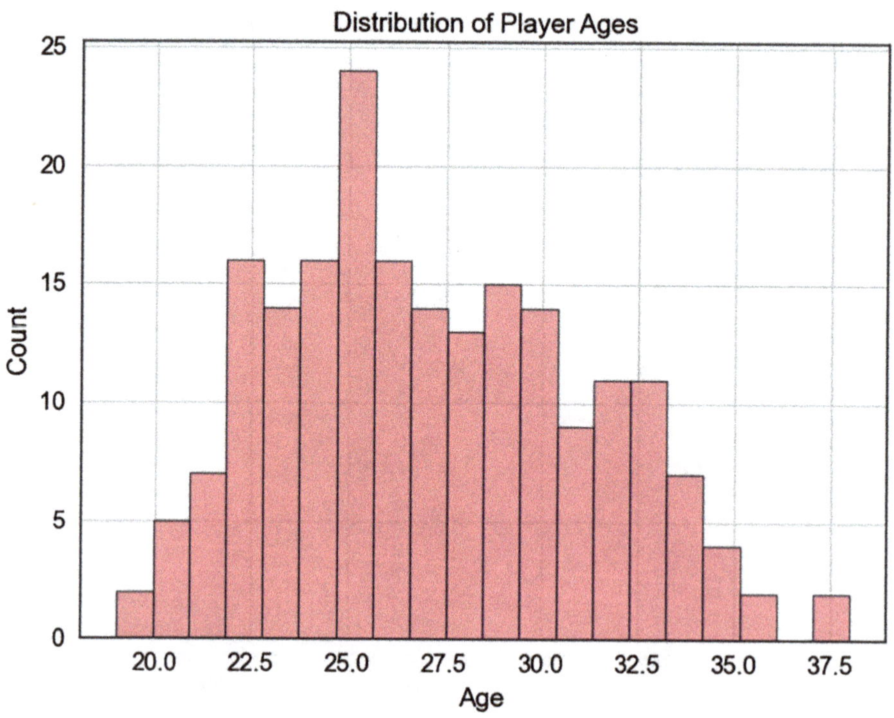

Fig. 3 Distribution of age

in this age group compared to other age groups. As the age deviates from its peak, the number of players gradually declines. This follows the standard pattern of a Gaussian distribution. The distribution of this product could have a substantial impact on teams or organizations seeking to attract or retain athletes. For instance, if the majority of the players fall within the age bracket of their mid-20s, it would be advantageous to concentrate on the recruitment or retention of players within that specific age group. Alternatively, if there is a deficiency of players within a specific age bracket, proactive measures could be used to specifically attract and enlist players within that range. Moreover, this data might be utilized to enhance training and development programs, as well as devise injury prevention and treatment techniques for players of varying age groups. In general, gaining knowledge about the dispersion of player ages can offer significant information for teams and organizations in the ISL.

11 Scatterplot Between Goals and Assists

The scatter plot depicting the correlation between assists and goals illustrates the connection between the number of goals achieved and the number of assists provided by players (Fig. 4). The clustering of data points within the range of 0–10 goals and 0–5 assists indicates that the majority of players own rather modest goal and assist figures. Nevertheless, there is a consistent upward trend in the number of assists as the number of goals increases, suggesting a positive correlation between goal-scoring ability and assist-making proficiency.

The two dashed lines, which represent the mean of assists and goals, demonstrate that the average number of assists made by players falls within the range of 3–4, while the average number of goals scored falls within the range of 4–6. Consequently, although the majority of players have modest goal and assist statistics, there exist a few players who excel and make substantial contributions to their team's achievements.

Businesses engaged in player scouting, sponsorship, and marketing could find this information highly beneficial. Businesses can utilize this data to discern individuals who demonstrate remarkable performance and possess the potential to ascend to stardom within the league. Additionally, this data can be utilized to strategically focus marketing campaigns on players who have achieved a significant number of goals and assists, as they are expected to have a greater appeal to fans and hence possess higher market value.

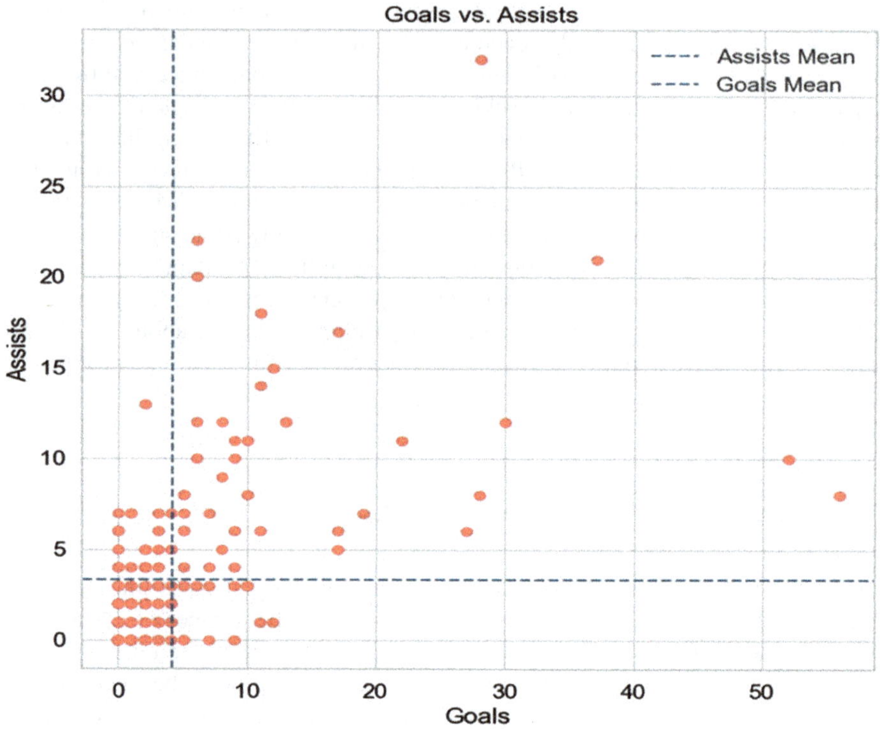

Fig. 4 Relationship between goals and assists

12 Results and Discussion

The objective of the analysis is to build a model that predicts the position of a soccer player based on their personal and performance attributes. The analysis used four different models, including multinomial logistic regression, random forest, support vector machine, and gradient boosting classifier, to achieve this objective. The accuracy of the models is one of the performance metrics used to evaluate the model's performance. The multinomial logistic regression and support vector machine models reached the highest accuracy of 0.82, while the gradient boosting classifier had the lowest accuracy of 0.69. The sensitivity and specificity of the models are additional performance metrics used to evaluate how well the model is predicting the outcomes. The sensitivity of a model refers to the proportion of true positives that are correctly identified, while the specificity of a model refers to the proportion of true negatives that are correctly identified. In this analysis, the support vector machine model achieved the highest sensitivity and specificity, indicating that it is a better predictor of the position of soccer players. The random forest model had the highest sensitivity but relatively low specificity, while the gradient boosting classifier had the lowest sensitivity and specificity among the models. Overall, the

Table 2 Model performance

Models	Accuracy	Sensitivity	Specificity
Multinomial logistic regression	0.81	0.85	0.92
Random forest	0.73	100	0.62
Support vector machine	0.81	0.97	0.97
Gradient boosting classifier	0.68	0.91	0.78

multinomial logistic regression and support vector machine models performed well in predicting the position of soccer players, based on their personal and performance attributes (Table 2).

13 Conclusion

In conclusion, the exploration of predicting player positions in the Indian Super League (ISL) through the lens of personal and performance attributes unfolds as a multifaceted journey within the broader landscape of football analytics. Drawing inspiration from a comprehensive literature review, which encompasses traditional spatial analyses, the influence of physical attributes, performance metrics, playing styles, and the integration of advanced machine-learning techniques, this chapter positions itself at the intersection of tradition and technological innovation in football analysis. As the ISL continues to carve its niche in the global football landscape, the synthesis of diverse insights from these studies informs a nuanced exploration into predicting player positions. The integration of machine-learning techniques and the analysis of personal and performance attributes within the unique ISL context offer potential avenues for groundbreaking findings. Through the analysis, it was found that the impact of personal attributes very specifically factors such as age and its role in player performance and prediction shows the vibrant nature of ISL. Also, it can be noted that the classification is achieved with more accuracy by non-ensemble methods like SVM and multinomial logistic regression. Also very surprisingly, the exploratory data analysis revealed that the performance is possible irrespective of the country, whereas the general belief in the football fraternity is performance has a significance with the country they belong to. The chapter not only contributes to the growing body of football analytics literature but also addresses a significant gap in understanding player positions specific to the Indian football context. In essence, this chapter serves as a bridge between tradition and innovation, leveraging the rich tapestry of football analytics to shed light on the dynamic nature of player roles in the Indian Super League. As the journey continues, the synthesis of these diverse studies will undoubtedly contribute to the broader discourse on the evolving landscape of football analysis and provide valuable insights for coaches, analysts, and enthusiasts alike.

References

Barraclough, D., Till, K., Kerr, R., & Emmonds, S. (2022). Talent identification in youth soccer: A systematic review. *Journal of Sports Sciences, 40*(1), 1–11. https://doi.org/10.1080/02640414.2021.1974298

Bloomfield, J., Polman, R., & O'Donoghue, P. (2019). Physical attributes that influence the playing position of elite soccer players. *Journal of Sports Sciences, 37*(4), 405–411.

Bradley, P. S., Carling, C., Gomez Diaz, A., Hood, P., Barnes, C., Ade, J., & Boddy, M. (2019). The effect of playing formation on high-intensity running and technical profiles in English FA Premier League soccer matches. *Journal of Sports Sciences, 37*(24), 2709–2716.

Buekers, M., Borry, P., & Rowe, P. (2015). Talent in sports. Some reflections about the search for future champions. *Movement & Sport Sciences, 88*(2), 3. https://doi.org/10.3917/sm.088.0003

Chatterjee, T. (2018). The state and future of football in India: A sociological perspective. *International Journal of Research in Sociology and Social Anthropology, 4*(2), 11–23.

Garris, M., & Wilkes, B. (2017). Soccernomics: Salaries for World Cup soccer athletes. *International Journal of the Academic Business World, 11*(2), 103–110.

Gonçalves, C. E. B., Rama, L. M. L., & Figueiredo, A. B. (2012). Talent identification and specialization in sport: An overview of some unanswered questions. *International Journal of Sports Physiology and Performance, 7*(4), 390–393. https://doi.org/10.1123/ijspp.7.4.390

Jamil, M., Phatak, V., Mehta, S., Beato, M., Memmert, D., & Connor, J. D. (2021). Predicting match outcome in football: The impact of quality-related factors, team rankings, and gender differences. *International Journal of Sports Science & Coaching, 16*(1), 106–118. https://doi.org/10.1177/1747954120944051

Lago-Peñas, C., Lago-Ballesteros, J., Dellal, A., & Gómez, M. (2016). Game-related statistics discriminated against winning, drawing, and losing teams from the Spanish soccer league. *Journal of Sports Science and Medicine, 15*(3), 418–425.

Leite, N., Calvo, A. L., Cumming, S., Gonçalves, B., & Calleja-Gonzalez, J. (2021). Editorial: Talent identification and development in sports performance. *Frontiers in Sports and Active Living, 3*, 729167. https://doi.org/10.3389/fspor.2021.729167

Link, D., & Lang, S. (2017). Advanced match analysis in football: Challenges and prospects. *International Journal of Sports Science & Coaching, 12*(4), 520–531.

Liu, Y., Bialkowski, A., Hingorani, H., & Ignjatovic, A. (2020). Predicting football player positions: A machine learning approach. *International Journal of Computer Science in Sport, 19*(1), 46–64.

Pappalardo, L., Cintia, P., Rossi, A., Massucco, E., Ferragina, P., Rossetti, G., et al. (2019). A public data set of spatio-temporal match events in soccer competitions. *Scientific Data, 6*(1), 1–15.

Reep, C., & Benjamin, B. (1968). Skill and chance in association football. *Journal of the Royal Statistical Society: Series A (General), 131*(4), 581–585.

Reep, C., Reilly, T., & Pollard, R. (2008). Case studies of factors influencing match performance in English FA Premier League soccer. *Journal of Sports Sciences, 26*(6), 643–650.

Sports Marketing Analytics: Digital Evolution in Shaping the Sports Industry

Sunanda Vincent Jaiwant and Joseph Varghese Kureethara

1 Introduction: Sports Marketing

Sports marketing analytics is the process of using data analysis to measure and optimize the effectiveness of marketing campaigns and strategies within the sports industry. It involves collecting and analysing data from various sources, such as social media, ticket sales, merchandise sales, television ratings, and fan engagement, to better understand consumer behaviour and preferences. Sports marketers can use analytics to identify trends and patterns in consumer behaviour and to measure the effectiveness of different marketing campaigns and strategies. By analysing this data, they can make more informed decisions about how to allocate marketing resources and improve the fan experience. Analytics have been used increasingly throughout the sport industry. The concepts of analytics and big data are frequently used in the sport industry and emphasized in numerous media outlets making analytics popular and important for the students of sports management for their future career (Mumcu & Fried, 2017). The potential of big data (BD) for sports marketing is immense, and there is a need for new data collection methods and analytical techniques. BD can provide unique insights for sports marketing scholars, helping to solve complex problems and make informed decisions. However, there are also many challenges in using BD in sports marketing, such as issues with data representation and algorithmic analysis. Since the potential of BD in sports marketing is promising, the scholars and practitioners should embrace its transformative

S. V. Jaiwant (✉)
School of Business and Management, CHRIST (Deemed to be University), Bangalore, Karnataka, India
e-mail: sunanda.vincent@christuniversity.in

J. V. Kureethara
CHRIST (Deemed to be University), Bangalore, Karnataka, India
e-mail: frjoseph@christuniversity.in

A. Mansurali et al. (eds.), *Sports Analytics*,
https://doi.org/10.1007/978-3-031-63573-1_13

potential. There should be more interdisciplinary research to fully explore the possibilities of BD in this field (Mamo et al., 2021).

The ease and thoroughness of data collection enabled by technology advancements has led to the emergence of machine learning and data analytics. This helps sports companies with their marketing campaigns to expand their fan base and increase merchandise sales. It also allows them to gather sponsors and run simulations of upcoming games. Through data analysis, sports companies may assess player performance and identify which recruits are necessary to increase team productivity. It evaluates the opponent's strong and weak spots to aid trainers in selecting the optimal tactic. Making use of data has the potential to increase revenue, reduce operating costs, and guarantee strong investment returns. The work of a sports data analyst entails gathering and analysing sports data, as well as informing players, trainers, and team leaders who utilize this information to arrive at judgments prior to or during sporting events. There are two groups that these responsibilities fall under. The initial application is in company operations, in which information is employed to raise the performance of a team or sports group's efficiency. This study aims to identify the concept of sports marketing analytics. The chapter will explore the different sports marketing analytics techniques which can be used in the field of sports marketing. The study will also make an effort to determine the advantages and difficulties of applying analytics to sports marketing. The notion and function of on-field and off-field metrics are the main topics of this study.

2 Concept of Sports Marketing Analytics

Sports around the world are a multi-billion-dollar industry. This vast industry owes its success to fan engagement and loyalty due to the data that fuels it. It is very important to appeal to the viewers of the sport which can be facilitated by gathering, analysing, and presenting data in an understandable manner. While it is essential to provide the viewers with a data-rich experience, the sports analytics teams are mindful to include graphics and visuals to avoid overwhelming the fans. Broadcasting of statistics, when the sport is live, and social media pages have proven to be crucial in ensuring excitement and interaction amongst fans. Currently, these are a couple of the most predominant ways of marketing that are being utilized by numerous sports brands and companies. Taking Formula1 as an example, platforms like F1 TV are providing a wholesome experience for its fans through its predictive analysis and detailed data points of the drivers throughout the race. This is giving the viewers a better understanding of the sport and therefore results in greater retention. However, for a long time, this global sport did not directly interact with its hardcore fans and viewers. F1 faced a major setback as the season entered the twenty-first century. There was very little connection with the ever-growing digital world and the fan bases were very weak. It was not until recently that the importance of digital presence as a marketing strategy was realized and implemented. Media coverage of the sport began to increase as it did its presence on several social media platforms.

The recent Netflix documentary 'Drive to Survive' revived the popularity and excitement of F1 amongst its fans and others unaware of the sport. Many viewers became eager to learn about F1 and began following the races closely post the release of the documentary. Post the COVID pandemic, viewership and fan engagement has multiplied manifold. Fans could analyse the data themselves, compare the several teams and drivers, and interpret their performances. This fan experience enhancement was possible largely because of the marketing strategy that was adopted. They focused on utilizing decades of data and ongoing race statistics which were presented to its fans regularly on social media. Not only was this data useful in attracting current fans and potential ones, but it also assisted the F1 marketing analytics team. They were now able to analyse present viewer preferences and viewing patterns, identify potential consumers of the content, and personalize the race experience online to make it consumable by all categories of fans. Similar to any other business, sports can succeed over the long haul only when it is managed well. Many sports organizations are substantially investing in the hiring of sports data scientists to conduct unbiased analyses of on- and off-field data to acquire a competitive edge. Data analytics and statistics are essential component that contributes to the growth of the industry. It will ensure transparency to all its stakeholders and provides an accurate picture of the trend patterns of the industry.

3 The Rise of Sports Analytics

Sports and statistical information were always closely associated. Historical data, like the scores in a game, has been used by audio announcers to provide context for their commentary. Newspapers print box scores, and sports cards feature a player's career statistics. Executives and trainers evaluate players based on a variety of statistical data, including yards thrown, batting average, and indications, depending on the sport and an emotional response. In addition to these research, further examination of data is typically resisted by both trainers and sportsmen. But early statisticians like Bill James began using data in the 1980s to challenge those illogical assumptions. James published the Sabermetrics statistical technique he created to rank baseball players in a book titled *The Bill James Historical Baseball Abstract*. He created mathematical formulas such as 'runs created' that predicted the number of runs a baseball team would likely score based on their offensive statistics. In accordance with the Society for American Baseball Studies, it was his first attempt at a strategy for helping executives enhance their teams by critically assessing players. The discipline of sports analytics took off when Oakland Athletics general manager Billy Beane used statistics to put together a team of less well-known players that nearly won a World Series in 2002. His 'Moneyball' strategy of statistically maximizing a team gained traction rapidly and became the accepted manner for managing a squad amongst rival teams. Since then, analytics have evolved uniquely in each major sport, with clubs employing data scientists and searching for fresh approaches to impartially assess players and obtain a competitive edge. According to a Research and Markets research from 2021, the global sports analytics market is expected to grow to $3.4 billion by 2028. Thanks to tracking software and computer learning, sports analytics has advanced to a new level. Companies such as Genius Sports can create statistical analysis from film footage to help coaches, enhance their play-calling during games, or generate postgame learnings. Businesses use cameras and machine learning algorithms to track player movements, ball speeds, and spin rates—factors that are commonly considered by teams and broadcasters when making decisions (Vandeput, 2021). For instance, it's common to see baseball players utilizing iPads to analyse information like pitching distribution while making changes mid-game. Sports fan consumption has also changed as a result of analytics. Supporters have access to website like FiveThirtyEight to learn more about data-based news about sports and the probabilities of their preferred team winning the title. In football or baseball, broadcast commentators frequently find out a player's breakaway speed or launch angle after a home run. Even when projecting the top players in fantasy sports, it is a mainstay. It took a while for sports analytics to advance from the sidelines to the front, and it doesn't seem like it will be leaving anytime soon.

The global sports analytics market is expected to grow at a compound annual growth rate (CAGR) of 22.3% from 2023 to 2030, from an estimated value of USD 2.73 billion in 2022. The demand for sports analytics solutions is anticipated to rise in the next years as on-field data becomes more crucial for analysing how teams perform, tracking each player, keeping records of injuries, and optimizing the workout plans of individual players. In order to create strategies, choose teams, and choose player line-ups for the game, team management have begun comparing off-field and on-field statistics of individual players. Administrators' greater reliance on real-time data to plan training sessions and implement strategic changes in games is anticipated to fuel additional industry expansion.

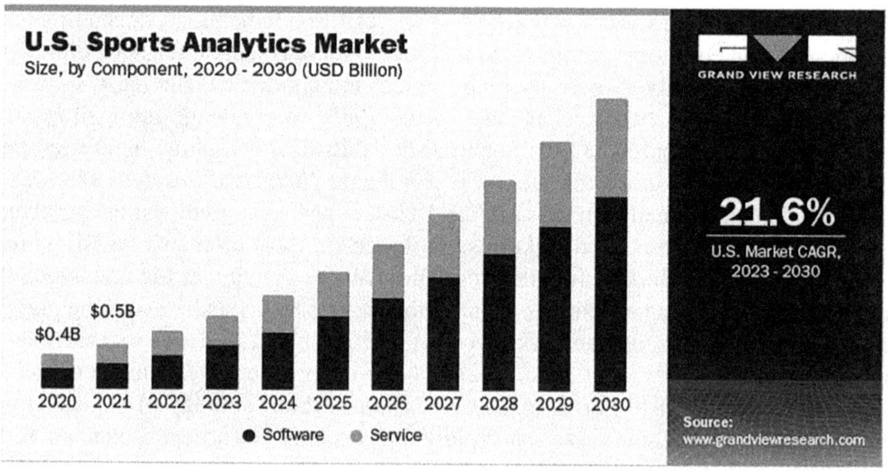

Source: GVR (n.d.)

The market for sports analytics is becoming increasingly more in demand as a result of ongoing technical improvements. Managers create game strategies based on statistics and data in order to succeed in-game through data analytics. These cutting-edge analytical solutions are essential for developing in-game strategies because they give management, stakeholders, coaches, and players access to solutions that produce insightful data on a variety of team and player-level metrics. In order to develop a more effective strategy, these technologies assist managers in understanding data on weather and pitch conditions, as well as data about players (GVR, n.d.).

Numerous emerging and well-established analytics firms are looking for ways to penetrate the sports industry and launch significant campaigns. For example, SIS, a well-known provider of sports data and analytics, joined with ReSpo.Vision in March 2023. ReDpo.Vision has agreed to give SIS information on sophisticated football analytics based on broadcast tracking technologies. The businesses will also introduce new tools to automate data collecting, monitoring, and the creation of innovative statistics for football (GVR, n.d.).

In the past, a game's outcome was purely determined by the players' skills. However, big data, AI, and machine learning (ML) technology advancements have caused a change in sports analytics trends. The machine learning techniques of this software examine factors like accomplishments at both domestic and international levels, competitors, and tournament circumstances and suggest the most effective performers for particular circumstances. Support vector machines and deep neural networks also assist administrators and trainers in figuring out combination and success possibilities (GVR, n.d.).

The clustering technique, on the contrary hand, focuses on specific categories to assist in growing the base of supporters using supporter management analysis. Big data analytics boosts team productivity and increases earnings from merchandise sales, sponsorships, media rights, and admission fees. The analytical data is frequently utilized in virtual gaming applications to display players' data and specific players when choosing players to get financial prizes. Virtual gaming's growing adoption of sports analytics tools is anticipated to have a favourable effect on the industry as a whole (GVR, n.d.).

4 On-Field and Off-Field Data Metrics

Sports analytics boosts the group's self-assurance and civic pride. The two primary subfields of sports analytics research are on-field analytics and off-field analytics. On-field analytics is a data-driven strategy for improving player and team performance, whereas off-field analytics examines the elements that assist the rights holders in maximizing corporate earnings. To comprehend and measure fan (human) behaviour, more sophisticated analysis techniques like sentiment analysis are now taking the place of conventional business intelligence (BI) methodologies. In order to forecast future events, highly complex deep learning and cognitive algorithms are used. Because of this, the market for sports analytics is expanding quickly.

On-field metrics in sports management refer to a collection of numerical statistics and information gathered during actual gameplay or on-field activities. Coaches, analysts, and team management use these indicators to evaluate player and team performance, come to wise conclusions, and create strategy. On-field metrics might differ greatly determined by the sports, but they are crucial for comprehending and raising efficiency. Having an earnings share of more than 61.0% in 2022, the on-field sector controlled the sports analytics market, primarily because of the growing demand for on-field analytical data such as players and team analysis, fitness evaluation, and video evaluation. Football, cricket, rugby, and swimming are just a few of the sports that have adopted on-field data analysis technologies in the past few years (GVR, n.d.).

For example, SAP SE provided TSG 1899 Hoffenheim, a German Bundesliga team, with the opportunity to use SAPs newest technologies for the club's digital transformation. Football administrators and trainers have benefited from the software's ability to dynamically visualize formations, plays, and strategy. Sports teams are anticipated to have an increased need for health assessments, player and team analyses, and video analysis over the projection period (GVR, n.d.).

5 Statistics Used in Sports to Measure the Player's Performance

Sport	Statistic	Definition
Baseball	Batting average	ratio of hits and number of at bats
	On-base percentage	times a player reaches base by hitting, walking, or by being hit by a pitch
	Slugging average	the ratio of number of bases earned and the number of at bats
	WHIP	Walks plus Hits allowed per Inning Pitched measure the number of baserunners the pitcher allows on hits and walks
Football	GAP ratings	Generalized Attacking Performance, introduced by Wheatcroft (2020), is a rating system for assessing the attacking and defensive capability of a team with respect to number of corners or shots in football.
	BA ratings	Bivariate attacking ratings is based on the parameter that minimises the mean absolute error (MAE) between estimated and observed match statistic.
Cricket: (Criclytics)	WASP	Winning and Scoring Prediction is a machine learning technique to predict final score of the first innings and probability of winning while chasing in second innings
	Win Probability Statistic	is a model that calculates the team's win probability in real-time using the team's historical data

AIM (2021)

Metrics from on-field data assist teams in making decisions about how to enhance game plans, diets, and other techniques for enhancing the efficiency of their athletes. Due to increased demands to boost productivity and protect athletes from injuries, it is anticipated that players and team analysis and health evaluation would experience significant development across the projection period. This information can be used in virtual sports games and wagering by sports fans. Throughout the course of the forecast time frame, this particular market segment is anticipated to grow at a considerable CAGR due to the growing desire amongst sports enthusiasts to comprehend sports fundamentals (GVR, n.d.).

The business aspect of sports analytics is covered by off-field analytics. In other words, off-field sports analytics is the act of looking at information that is not directly associated with the game. Numerous sources, such as social media, ticket sales, financial records, and more, can provide this information. Everything in the sports industry that doesn't involve on-field performance is referred to as off-field analytics. It emphasizes more commercial elements like sales and ticket prices. Sports organizations can utilize this information to expand their operations and improve profitability.

The use of data to uncover patterns and insights, for instance, can help to increase ticket and merchandise sales as well as fan engagement. This is known as off-field analytics.

Sports management relies heavily on on-field measurements to help teams and coaches analyse how players perform, adjust plans, and arrive at data-driven decisions. Technology and data analytics developments have rendered it feasible to gather and analyse a wide range of on-field measures, offering insightful information about the nuances of sporting performance. From 2023 to 2030, the off-field market is anticipated to grow at a CAGR of roughly 20.0%. Off-field analytics improves a team's or club's business operations and operational effectiveness. Different stakeholders, fantasy games, and betting applications all heavily rely on off-field analytics tools to help them make profitable selections (Eric, 2023).

Some specific areas where sports marketing analytics can be applied include:

Fan Engagement With the help of big data and sports marketing analytics, sports organizations can analyse their fans' preferences, interests, and behaviour patterns to create personalized engagement strategies. They can use social media, mobile apps, and other channels to deliver customized messages, promotions, and experiences to fans. By doing so, sports organizations can increase fan loyalty, drive ticket sales, and generate additional revenue streams.

Player Performance Analysis Sports organizations can use big data and sports marketing analytics to analyse player performance data to identify trends and patterns. This information can be used to improve training programs, scouting efforts, and player development initiatives. By doing so, sports organizations can identify and develop top-performing players, ultimately improving their team's performance and increasing their chances of winning.

Sponsorship Management Sports organizations can use big data and sports marketing analytics to analyse sponsorship data to identify the most effective sponsorships and marketing campaigns. This information can be used to negotiate better deals with sponsors and to create targeted marketing campaigns that resonate with fans. By doing so, sports organizations can generate more revenue from sponsorships and other marketing initiatives.

Ticket Sales Optimization Sports organizations can use big data and sports marketing analytics to analyse ticket sales data to identify trends and patterns. This information can be used to develop more effective pricing strategies, targeted marketing campaigns, and better customer service initiatives. By doing so, sports organizations can increase ticket sales and generate more revenue (Smolic, 2022; Urwin, 2022).

Facility Management Sports organizations can use big data and sports marketing analytics to analyse facility management data to identify opportunities to improve facility utilization, reduce operating costs, and improve the overall fan experience. This information can be used to optimize facility design, improve maintenance schedules, and create targeted marketing campaigns that promote facility features and amenities. By doing so, sports organizations can increase revenue, reduce costs, and improve the overall fan experience.

The applications of big data and sports marketing analytics are vast and varied. By leveraging these technologies effectively, sports organizations can increase revenue, reduce costs, improve fan engagement, and ultimately succeed over the long haul.

6 Social Media's Impact on Sports

Over many years, the dynamics of different sectors have been changing as a result of media in general. And the sports sector is no different; it started with live broadcast TV, then moved to cable, then OTT subscription, and is now using social media. Over the years of COVID seclusion, the emergence of social has been particularly swift. It's simple to comprehend at its foundation. A one-way push of information, broadcast television. Sports fans may interact, respond, and feel more connected to their favourite teams and stars because to the two-way communication that social media facilitates. The influence of social media on sports and the way fans watch sports have increased dramatically. Fans no longer have to wait for the morning newspaper for the most recent information. The general public is informed immediately as transactions and agreements take place. Using social networking sites gave followers and the general public fast accessibility to the latest developments and data comments and responses from individuals all across the world. Since current research indicate relatively independent or strengthening relationships between social media use and sports involvement, rising social media activity is frequently explained as replacing participation in sports. Online social networks like Facebook and Instagram are examples of specialized social media platforms that offer a wide range of sports-related content, including visual representations of sports, the sharing of training schedules and tips for building muscle, as well as discussions about sporting goods. However, in scientific contexts, the influence of these materials on sporting activities has been generally ignored.

Fans of sports increasingly decide to follow their preferred teams on social media. Sports marketers believe that now is the perfect moment for all sports leagues, teams, national organizations, and competitions that haven't yet embraced the rapidly growing new social networking to do so. 41% of global sports fans watch live matches on digital platforms. 32% of all sports fans, or 43% of Gen Z aficionados, use social media networks when watching live sports. These and other statistics demonstrate the enormous and constantly expanding influence that social media has on sports and spectators. The contemporary sports world has grown increasingly dependent on social networking. As more people spend more time online, fans utilize social media to talk about and participate with topics related to sports and content. Businesses are also coming up with creative ways to help their sports teams and promote sports conversation. Viewers these days more than just watching sports want to engage lively with tournaments and matches by sharing memes, and engaging in trash-talking. They interact with them directly (Gerrard, 2016).

The era of sports fans only being able to interact with their favourite sports teams or athletes by attending live games or gathering trading cards is a thing of the past. Sports fans now have the ability to follow their preferred teams on social media and interact with athletes, coaches, and sports analysts (Charts, 2021).

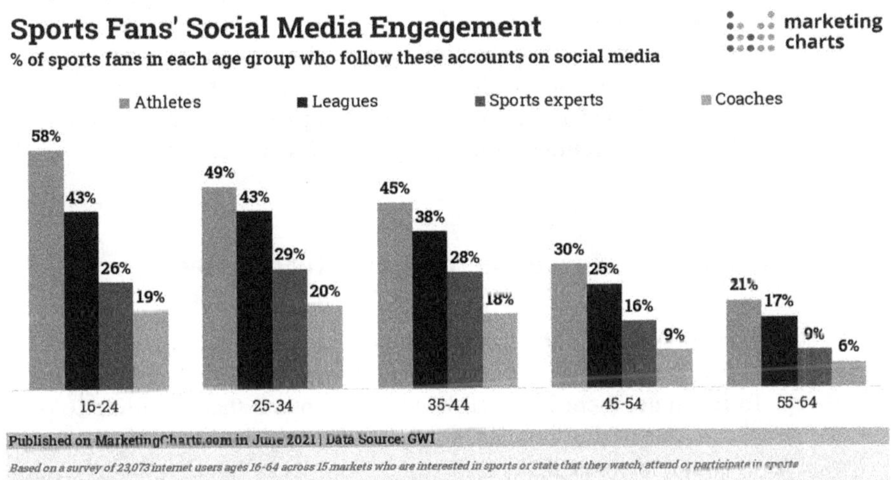

Source: Charts (2021)

Younger fans are more inclined to connect with sports figures and teams on social media, according to a study of Internet-using sports enthusiasts (ages 16–64) in 15 international markets. However, fans of all ages are following sports figures and teams to some extent. This is consistent with recent research demonstrating that younger individuals are more inclined to utilize social media and follow influencers or brands there. Such younger supporters also tend to be more likely to be sports aficionados. Approximately 58% of Internet users between the ages of 16 and 24 and 49% of those between the ages of 25 and 34 report following athletes. The proportions of people who following leagues (each at 43%), sports analysts (each at 26 and 29%), and trainers (each at 19 and 20%) are in comparison. The majority of sports enthusiasts (52%) choose YouTube, followed by Facebook (47%) and Instagram (43%) as their preferred sports-watching sites. Additionally, younger fans are more likely than older ones to watch sporting events online, despite the fact that the majority of spectators of all ages claim to watch athletic events on television. For instance, more than three-quarters of Millennials (74%) claim to watch sports online, compared to less than half (46%) of Baby Boomers. In addition to social media, fantasy leagues are another pastime for sports lovers. Cricket (35%), American football (22%), and soccer enthusiasts (21%) are the sports fans who participate the most. Fans of American football, motorsports, soccer, and tennis claim they were most aware of sponsorship on shirts, followed by LED boards around the field or stadium, particularly regarding advertising (Charts, 2021).

Another notable advantage of social media for sports organizations is its capacity to interact directly with fans and make them feel like they are a part of the league. Especially uncommitted viewers—those viewers who don't watch many broadcasts and may not be dedicated to any particular teams or players. The majority of people's Internet time is spent browsing to locate content that piques their interest, and social media is a terrific way to keep them informed about the sport, club, or league. Engaging on social media is a great way for sports organizations to reach a wider audience. Many sports leagues use their social media platforms to promote their events and activities and interact with fans through content about sports. These organizations use social media sites to publish line-ups for new games, promote players, and publicize forthcoming games (Alamar & Mehrotra, 2019).

7 Different Sports Marketing Analytics Techniques

1. Target audience—Via the data that is gathered and analysed, a specific set of customers must be targeted to streamline the marketing activities and product delivery. To illustrate, there is a well-known shoe brand that is willing to promote a particular line of shoe products for the occasion of the upcoming Women's Day. The story and idea that the brand wants to convey is that all women regardless of their age, size, or shape would feel empowered and ready to take on any challenge when they wear the shoe. The idea originated with the changing mindsets and lifestyles of consumers. Through this campaign, women were encouraged to push themselves and aim at physical fitness. Given the rising awareness of leading a healthy lifestyle, this campaign would be launched and specifically targeted at women.
2. Social Media—This is an extremely vital tool in the present day and age that has proven useful time and again to understand consumer tastes and preferences. The exposure it provides to the respective sports and fitness enthusiasts helps in boosting brand awareness. A tremendous increase in content consumption was observed during and post-pandemic years. The ability of fan reactions and engagement is the main motivator for consumers to follow the content and for marketers to publish such data online. FIFA, NBA, and F1 amongst others have

extensively utilized the online space in recent years to provide analytics of the respective sports to the fans, thereby ensuring greater interest.

3. Engaging Content—When fans and viewers of sports are given material that catches their attention, there is a greater possibility of loyalty and retention. They do not feel too overwhelmed with the content when the facts and figures of the sports are delivered in a balanced manner. One of the most popular ways of achieving this is by involving famous personalities and sports icons in advertisements, collaborations, and other promotional activities. When viewers feel like they are represented through the celebrities they admire, they are more likely to build trust and association with the particular brand.
4. Seasonal Sports—It is important to create a pre-game buzz amongst viewers and fans to garner their attention and enthusiasm. For instance, during the Cricket World Cup, there is a wave of excitement, anxiety, and eagerness amongst cricket fans all over the world. This is a perfect time for data to be collected and analysed. Sports-related content is delivered during commercial breaks, fan surveys are recorded, and there is much more engagement in the whole of the industry than during the off-season. This is a great opportunity for sports brands to exploit and fan commitments are guaranteed during the sports seasons.

8 Benefits and Challenges of Using Analytics in Sports Marketing

8.1 Benefits

Technological advances have made it feasible to access seemingly unimaginable amounts of data. Teams involved in sports can benefit from this knowledge that is accessible. The sports industry uses sports analysis to increase revenue, improve performance on both individual and team levels, lower injury rates, and make many other improvements. All of this information is a great resource, but it is meaningless until someone evaluates and takes into account its possible uses. Sports analysts are in high demand at the moment since many teams are dedicating entire departments to data analysis in an attempt to become the best league program. In other words, teams in sports use analytics to gain an advantage over their opponents (Kovalchik, 2018). Technology breakthroughs have made it possible to access an amount of data that is apparently inconceivable. This information is available, and sports teams can take advantage of it. Sports analysis is used by the sports industry to boost profits, better individual and team performance, and reduce injuries and for many other improvements. Even while all of this information is an excellent resource, it is useless without someone to assess and consider its potential applications. As many clubs create whole divisions only for analysing data in an effort to establish themselves as the greatest league program, sports analysts are in great demand right now. In other words, teams in sports use analytics to gain an advantage over their opponents (Olsofka, 2018):

- Consumers—presenting meaningful and understandable data and analytics of the sport is highly essential to ensure the growth of fan bases. It creates an enhanced experience for the fans and viewers and increases their interest in the sport.
- Marketing analysts—to gain a competitive advantage, sports analysts are a core part of the marketing team at any sports organization in the present day. As technology advances, analysts must be aware of the various developments in data communication and collection. It would be useful to utilize data on fans' online consumption patterns to update their marketing strategies.
- Wearable technology—this could assist in collecting data from the players of the sport. Their movements and data points can be analysed and communicated to fans. As such resources are growing rapidly, they are proving to be one of the best resources for analysts to carry out their activities and produce accurate data.
- Increase in revenue—data regarding viewer footfalls and revenue generated from each game may be recorded. This data is analysed by the sports analytics team and passed on to the marketers who can then make decisions on the deterioration or improvement of the turnout. Reasons for such turnouts and sales can be determined and analysed (Bajaj et al., 2020). This will help in multiplying the revenue for the company and ensure long-term growth.

8.2 Challenges

Despite its swift development, sports analytics still has a lot of obstacles to overcome. Sports analytics detractors argue that certain elements, such as players jumping in a game, deceiving opponent players, and players screaming, are impossible for statistics to capture. They contend that only people are capable of capturing and processing these kinds of information. Nevertheless, analytics continue to operate with these unorganized information types to some degree. Events like these are recorded and processed using text analytics algorithms, transforming unorganized facts into conventional organized data with rows and columns. To estimate word frequency and produce insights, models based on rules or algorithms are used. More data from other sources can be gathered in order to increase the effectiveness of these models.

- Data overload—the content delivered to the fans and viewers can easily be deemed as complex and distractive. There is a need to carefully curate necessary data and analytics and provide valuable statistics. The end consumer should be able to easily grasp and interpret the data presented. This is a very crucial step for sports marketing analysts to avoid a decrease in fan bases.
- Data privacy—this is a chief concern involving several regulatory authorities which is not unique to the sports industry. Since online hacking and Internet safety are not guaranteed wholly, protecting the interests of fans is a very important function. Consumers (fans) must be asked for consent and be aware of the reason for collecting data.

- Loss of data—maintaining a safe and secure line for the passage of data is a challenging task. A huge network of intermediaries is involved including remote and mobile assets which makes it difficult to monitor. Due to its complex nature, the protection of transmission and collection of data is most important. The sports organizations' Intellectual Properties (IP) will be jeopardized in fiercely competitive industries such as this.

9 Next in Sports Analytics

The sports sector recently experienced an unprecedented development because of sports analytics, yet there is still plenty to be done. With the advancement of technology and wearables, the time is not far off when analytics will evaluate a player's mental and emotional makeup and how it connects to their performance on the field (KU, n.d.). The field of sports analytics is very broad and will continue to develop throughout the years. Additionally, the sale of tickets online present fresh chances to employ analytics and data to enhance the client experience. It was difficult to determine clients back when tickets were printed on paper. Nowadays, one can enter with one's phone, and the majority of venues are now entirely digital. That is a fantastic illustration of how data and technology are used to identify customers. As a result of evolving technology, data, and analytics, the ability to tailor the experience, deal with crisis in the case of a security breach, or manage operations within a venue increases significantly. Sports organizations use analytics to help them make decisions about expanding into other markets. They utilize data science to choose which markets to target, how much money to invest, and how much it will cost to acquire a customer compared to how much that customer will be worth over their lifetime (Apostolou & Tjortjis, 2019).

Big data and analytics can be used to improve athlete safety. The Wharton Neuroscience Initiative allows specialists to collect precise data on the brain and monitor a player's brain function over time. In the future, real-time analytics may identify injuries while players were on the field, particularly if a player was not performing optimally. In general, analytics will advance the sports betting sector, particularly in the two highlighted areas. First is in the evaluation of risks, wherein innovative goods that improve the sports betting experience will be helped to reach consumers through technology. Another factor pertains to the speed with which technology is developing, for example, the metaverse and the way it materially affects customer experiences and customer sports betting behaviour (AltexSoft, 2022). The sports industry might also benefit from studying other sectors, such as financial services for the convenience of payments or Amazon for streamlining the consumer experience. Consumers ultimately desire a user-friendly betting platform. People desire to have a quick and simple way to deposit money and get their profits out.

10 Conclusion

Sports analytics has become a rapidly growing subject of study due in part to the real-world success depicted in the best-selling book and movie *Moneyball.* On the field, court, and ice as well as in people's living rooms with fantasy sports players and online sports gambling, analysis of team and player performance data has continued to change the sports industry. Sports players, coaches, experts, and marketers learn how to create predictive models to foretell team and player performance using real data sets from the Major League Baseball (MLB), National Basketball Association (NBA), National Hockey League (NHL), English Premier League (EPL-soccer), and Indian Premier League (IPL-cricket). Additionally, one can learn how to use real statistical models to replicate Moneyball's success, how to predict categorical outcome variables in sporting events using the linear probability model (LPM), how teams gather and organize an athlete's performance data using wearable technology, and how to use machine learning in a sports analytics context. Data and analytics are being employed more and more to improve player safety while also maximizing fan experiences and team owners' ability to assess player value. There is a growing awareness about the potential use of analytics in identifying player ailments in real time and taking preventative measures. The sports sector must simultaneously guarantee the security and privacy of athlete and customer data. Although team management frequently utilize data analytics to assess player value, it is also being used in a variety of other contexts. In order to better understand sleeping habits and how both correlate to improved efficiency, sensors could be attached to athletes while they practise or rest. The data are important. The analytics will be more accurate the more detailed the data is. The collection of data is still in its infancy for several sports. Soccer does not have the same amount of data as basketball, football, MLB, etc.

Several trends that will likely emerge in the future are promising for sports marketing. Artificial intelligence (AI) and machine learning (ML)-based technological breakthroughs present prospects for improved customization and targeted marketing. By enabling secure and transparent transactions, blockchain technology and cryptocurrencies could revolutionize the sports sponsorship industry (Priyal, 2023). The fan experience will be improved by integrating e-commerce and fan gear. Fans will be able to develop a closer relationship with their favourite teams and sportsmen, thanks to online retail platforms and personalized merchandise, which also provides exclusive deals and limited-edition items. Sports marketing will keep on taking on the characteristics of the OTT (over-the-top) platforms and live streaming as they develop. Sports organizations and businesses may interact directly with fans by distributing material straight to their homes, providing them with distinctive and immersive experiences. The increase in revenue will also be fuelled by monetization possibilities through subscriptions and in-app payments.

References

AIM. (2021). A primer on sports analytics: A new dimension of sports. . Retrieved from https://analyticsindiamag.com/a-primer-on-sports-analytics-a-new-dimension-of-sports/

Alamar, B., & Mehrotra, V. (2019, March). Beyond Moneyball: The future of sports analytics. *Analytics Magazine*. Retrieved from http://analytics-magazine.org/beyond-moneyball-the-future-of-sports-analytics/

AltexSoft. (2022). *Demand forecasting methods: Using machine learning to see the future of sales*. Retrieved from https://www.altexsoft.com/blog/demand-forecasting-methods-using-machine-learning/

Apostolou, K., & Tjortjis, C. (2019). Sports Analytics algorithms for performance prediction. In *10th International Conference on Information, Intelligence, Systems and Applications (IISA)* (pp. 1–4). IEEE. https://doi.org/10.1109/IISA.2019.8900754

Bajaj, P., et al. (2020). Sales predicion using machine learning algorithms. *International Research Journal of Engineering and Technology, 7*(6), 3619–3625. e-ISSN: 2395-0056.

Charts, M. (2021). *Sports fans display strong social engagement*. Retrieved from https://www.marketingcharts.com/industries/sports-industries-117340

Eric, J. (2023). *Sports analytics: Complete intro to sports data analytics*. Datarundown. Retrieved from https://datarundown.com/sports-analytics-intro/

Gerrard, B. (2016, May). Moneyball and the role of sports analytics: A decision-theoretic perspective. In *North American Society for Sport Management Conference* (pp. 108–109). NASSM

GVR. (n.d.). *Sports analytics market size, share & trends analysis report* (Report ID: GVR-4-68038-128-3). Grand View Research.

Kovalchik, S. (2018). The game insight group: A model for academic-industry partnerships for sports statistics innovation. *Quality Engineering, 31*, 23–38. https://doi.org/10.1080/08982112.2018.1519578

KU. (n.d.). *Sports marketing trends and best practices: Reaching fans in a digital age*. Retrieved from https://onlinesportmanagement.ku.edu/community/sports-marketing-trends

Mamo, Y., Yiran, S., & Andrew, D. (2021). The transformative impact of big data applications in sport marketing: Current and future directions. *International Journal of Sports Marketing and Sponsorship, 23*, 594–611. https://doi.org/10.1108/IJSMS-03-2021-0073

Mumcu, C., & Fried, G. B. (2017). Analytics in sport marketing. *Sport Management Education Journal, 11*, 102–105.

Olsofka, M. (2018). *Why is data analytics so important in sports?* Retrieved from https://www.samford.edu/sports-analytics/fans/2018/Why-is-Data-Analytics-So-Important-in-Sports

Priyal. (2023, July 4). *The evolution of sports marketing*. Retrieved from https://www.themediaant.com/blog/the-evolution-of-sports-marketing/

Smolic, H. (2022, September 1). *Sales forecasting: How to apply machine learning*. Retrieved from https://graphite-note.com/machine-learning-sales-forecasting

Urwin, M. (2022, August). *How sports analytics are used today, by teams and fans*. Retrieved from https://builtin.com/big-data/big-data-companies-sports

Vandeput, N. (2021, October 11). *How to: Machine learning-driven demand forecasting*. Retrieved from https://towardsdatascience.com/how-to-machine-learning-driven-demand-forecasting-5d2fba237c19

Index

A. Mansurali et al. (eds.), *Sports Analytics*,
https://doi.org/10.1007/978-3-031-63573-1

www.ingramcontent.com/pod-product-compliance
Lightning Source LLC
Chambersburg PA
CBHW071701220925
32987CB00003B/220
* 9 7 8 3 0 3 1 6 3 5 7 2 4 *